Public/Private Partnerships
Innovation Strategies and
Policy Alternatives

Public/Private Partnerships
Innovation Strategies and
Policy Alternatives

by

Albert N. Link

 Springer

ISBN-13: 978-1-4419-4000-1

e-ISBN-10: 0-387-29775-8
e-ISBN-13: 978-0387-29775-0

Printed on acid-free paper.

Printed in the United States of America.

9 8 7 6 5 4 3 2 1

springeronline.com

for Carol, Jamie, and Kevin

CONTENTS

LIST OF TABLES

—

LIST OF FIGURES

ACKNOWLEDGEMENTS

My thanks to my many friends who offered comments and suggestions on earlier drafts of this manuscript. A special thanks to Todd Crawford, John Hardin, Jamie Link, Donald Siegel, and John Scott. And, of course, there is my wife, Carol, who was patient throughout the writing and re-writing and the proofing and re-proofing stages.

1 INTRODUCTION

$\mathcal{R}$esearch and development (R&D) leads to innovation and innovation to technological change. Technological change, in turn, is the primary driver of economic growth. Public/private partnerships leverage the efficiency of R&D and are thus a critical aspect of a nation's innovation system.

Public/private partnership is a term that is becoming more and more widely used in economics and in policy circles.[1] As is common in these and other disciplines, there are terms of art and terms of science; public/private partnership is a term of art without a precise, much less generally accepted, definition.

PUBLIC/PRIVATE PARTNERSHIPS

"Public," as the term public/private partnership is used within the context of this book, refers to any aspect of the innovation process—a term to be defined below—that involves the use of governmental resources, be they federal, state, or local in origin. "Private," refers to any aspect of the innovation process that involves the use of private sector resources, mostly firm-specific resources. And, resources are broadly defined to include all resources—financial resources, infrastructural resources, research resources, and the like—that affect the general environments in which innovation occurs. Finally, the term "partnership" refers to any and all innovation-related relationships, including but not limited to formal and informal collaborations in R&D.

---

[1] This discussion about the term "public/private partnership" draws in large part from a study funded by Link for the OECD Committee for Scientific and Technology Policy. The report was later published in an abbreviated form in Link (1999a).

The above definitions of "public" and "private" are straightforward, but some might pause over the definition of "partnership." Surprisingly, there is not a generally accepted definition for that term in the economics or policy literatures, especially with relevance to innovation. Coburn (1995, p. 1) used that term synonymously with cooperation by defining cooperative technology programs as:

> ... public-private initiatives involving government and industry—and often universities—that sponsor the development and the use of technology and improve practices to measurably benefit specific companies.

More narrowly, Link and Bauer (1989) defined research joint venture (RJV) partnerships as arrangements through which firms jointly acquire technical knowledge.

The National Research Council (Wessner 2003, p. 7) offered an explanation of a public/private partnership in terms of what it is and what it does:

> Public-private partnerships involving cooperative research and development among industry, government, and universities can play an instrumental role in introducing key new technologies to the market ... [Partnerships] often contribute to national missions in health, energy, the environment, and national defense and to the [N]ation's ability to capitalize on its R&D investments.

The definition set forth in this book follows in spirit from that used by the Council on Competitiveness (1996, p. 3):

> Partnerships are defined ... as cooperative arrangements engaging companies, universities, and government agencies and laboratories in various combinations to pool resources in pursuit of a shared R&D objective.

Based on the Council's definition, a public/private partnership is a relationship—either formal or informal among participants in the R&D process, or institutional—that involves the use of public and/or private resources be they financial, infrastructural, or research based.

The Council on Competitiveness's definition raises an important issue, namely: Why should public resources be used in partnership with private resources? More broadly: What is the economic rationale for public/private partnerships?

PUBLIC/PRIVATE PARTNERSHIP FRAMEWORK

The framework which defines the public/private partnership focus of this book can be described in terms of Table 1.1. The first column of the table describes the nature and scope of governmental involvement in a public/private partnership. Governmental involvement could be indirect or direct, and if direct there is then an explicit allocation of resources including financial, infrastructural, and research.

The second and third columns in the table relate to the economic objective of the public/private partnership. Of course, with any innovation-related activity there are spillovers of knowledge and thus economic objectives are multi-dimensional, but for illustrative purposes herein a single overriding economic objective is assumed. Broadly, the objectives are to leverage public R&D activity, or to leverage private R&D activity.

Each public/private partnership discussed in this book is mapped into the format illustrated by the template in Table 1.1.

Table 1.1. Taxonomy of Public/Private Partnerships

	Economic Objective	
Governmental Involvement	*Leverage Public R&D*	*Leverage Private R&D*
Indirect	...	...
Direct		
Financial Resources	...	...
Infrastructural Resources	...	...
Research Resources	...	...

Table 1.1 is only one categorical approach to public/private partnerships. Coburn (1995), for example, classified public/private partnerships in terms of the benefits and services that they offer to industry. Toward that end, he posited five functional categories:

(1) Technology Development: research and applications for new or enhanced industrial products and processes.

(2) Industrial Problem Solving: identifying and resolving firm-level industrial needs through technology and best-practice applications.
(3) Technology Financing: public capital or help in gaining access to private capital.
(4) Start-up Assistance: aid to new, small technology-based businesses.
(5) Teaming: help in forming strategic partnerships and alliances.

Alternatively, the Office of Technology Policy (1996) classified public/private partnerships in the United States along a time spectrum so as to illustrate and emphasize that public/private partnerships have evolved from a relationship wherein the government was merely a customer of private research to a relationship wherein the government is a partner in research. In other words, the Office of Technology Policy's taxonomy is one that stresses the evolution of the public role in partnerships. Specifically (Office of Technology Policy 1996, pp. 33-34):

> By the late 1980s, a new paradigm of technology policy had developed. In contrast to the enhanced spin-off programs—enhancements that made it easier for the private sector to commercialize the results of mission R&D—the government developed new public-private partnerships to develop and deploy advanced technologies. ... [T]hese new programs ... incorporate features that reflect increased influence from the private sector over project selection, management, and intellectual property ownership. Along with increased input, private sector partners also absorb a greater share of the costs, in some cases paying over half of the project cost.
> The new paradigm has several advantages for both government and the private sector. By treating the private sector as a partner in federal programs, government agencies can better incorporate feedback and focus programs. Moreover, the *private sector as partner* [emphasis added] approach allows the government to measure whether the programs are ultimately meeting their goals: increasing research efficiencies and effectiveness and developing and deploying new technologies.

Finally, the National Research Council (Wessner 2003, p. 8) categorizes public/private partnerships in terms of their contribution:

... to the development of industrial processes, products, and services that might not otherwise emerge spontaneously, and in this way help address government missions and generate greater public welfare.

According to the National Research Council (Wessner 2003, pp. 8-9), addressing government missions and generating public welfare is related to the following:

- Developing new technologies often require collective action, particularly in the case of high-spillover goods, where technology advances generate benefits beyond those that can be captured by innovating firms. Partnerships can be a means of encouraging the cooperation necessary for socially valuable information.
- New technologies often involve investments in combinations of technologies that may remain unexploited ... in companies or industries. Joint research [partnership] activities can facilitate the cooperation necessary to achieve the commercial potential of these technologies.
- Partnerships encourage firms to undertake socially beneficial R&D. The return on R&D investment, even for promising technologies, can be perceived to be too low when firms heavily discount distant income streams or when risks related to technical development and commercialization are seen as substantial.

OVERVIEW OF THE BOOK

The remainder of the book is divided into two sections. Chapters 2 through 5 provide an overall framework for the book, and Chapters 6 through 14 relate to specific U.S. public/private partnerships.

The framework chapters accomplish two goals. The first goal is to set forth an economic argument for the public's role—government's role—in innovation, in general, and in public/private partnerships, in particular. This is done in Chapters 2 and 3. The second goal is to place innovation and the innovation process within a broader economic model of technological change. This is done in Chapters 4 and 5.

The goals of the remaining chapters are to illustrate an aspect of U.S. innovation policy through a description of a number of public/private partnerships, and to evaluate their social impacts, to the extent possible, based on the extant literature. Chapters 6 through 10 deal with specific U.S. public/private partnerships and initiatives that leverage private-sector R&D. The partnerships include the U.S. patent system, tax incentives toward R&D, research collaborations including research joint ventures, and, as a specific institutional illustration, the Advanced Technology Program (ATP). Chapters 11 and 12 deal with partnerships that leverage public-sector R&D including the laboratory research at the National Institute of Standards and Technology (NIST, which also leverages private-sector R&D) and the Small Business Innovation Research Program (SBIR). Evaluation methods relevant to public/private partnerships are discussed in Chapter 13, and a concluding statement is in Chapter 14.

2 THE HISTORY OF PUBLIC/PRIVATE PARTNERSHIPS

The development of science, technology, and economic growth in the United States was greatly influenced by the scientific discoveries and university infrastructure within Europe during its colonial period. While it is difficult to pinpoint how or which specific elements of scientific and technical knowledge diffused across the Atlantic, certain milestone events can be dated and key individuals can be identified.

The background in this chapter, which draws on Unesco (1968) and National Science Board (2000), gives not only an appreciation for the role that science and technology resources have played in the development of the Nation, but also historical insights into the evolution of public/private partnerships in the United States.[1]

THE COLONIAL PERIOD

The first member of the Royal Society of London to immigrate to the Massachusetts Bay Colony was John Winthrop, Jr. in 1631, just a few years after the founding of the Colony. As a scientist, he is credited with establishing druggist shops and chemistry laboratories in the surrounding villages to meet the demand for medicine. According to Unesco (1968, p. 9), these ventures were "perhaps the first science based commercial enterprise of the New World."

Before the turn of the eighteenth century, colonists made noticeable advances toward what may be called a scientific society, organizing

---

[1] The original version of this chapter was set forth in Link (1999b), later expanded in Audretsch et al. (2002a), and then reproduced in book form as Feldman, Link, and Siegel (2002).

scientists who came from England and other European countries into communities that promoted scientific inquiry. In 1683, the Boston Philosophical Society was formed to advance knowledge in philosophy and natural history.

Benjamin Franklin formed the American Philosophical Society of Philadelphia in 1742 for the purpose of encouraging correspondence with colonists in all areas of science. This Society later merged with the Franklin-created American Society to promote what Franklin called "useful knowledge," and it still exists today. The combined society focused on making available advancements in agriculture and medicine to all individuals by sponsoring the first medical school in America (also supported by the Pennsylvania House of Representatives). Thus, Franklin's combined society was a hallmark of how public and private sector interests could work together for the common weal.

Influenced by the actions of Pennsylvania and later Massachusetts with regard to sponsorship of scientific institutions, the establishment of national universities for the promotion of science was first discussed at the Constitutional Convention in 1787. However, at that time, the founders of the Constitution believed educational and scientific activities should be independent of direct national governmental control. But, they felt that the national government should remain an influential force exerting its influence through indirect rather than direct means.

For example, Article I, Section 8, of the Constitution states:

> The Congress shall have the power ... To promote the progress of science and useful arts, by securing for limited times to authors and inventors the exclusive right to their respective writings and discoveries.

Soon thereafter, in 1790, Congress passed the first patent act.

Alexander Hamilton, in his role as Secretary of the Treasury, released on December 5, 1791 A Report on Manufacturers. Therein he advocated a direct role of the government in support of the Nation's manufacturing:

> The expediency of encouraging manufacturers in the United States, which was not long since deemed very questionable, appears at this time to be pretty generally admitted. The embarrassments, which have obstructed the progress of our external trade, have led to serious reflections on the necessity of enlarging the sphere of our domestic commerce; the restrictive regulations, which in

> foreign markets abridge the vent of the increasing surplus
> of our Agriculture produce, serve to beget an earnest
> desire, that a more extensive demand for that surplus may
> be created at home ...

Thomas Jefferson also championed a more direct role for the government in the area of science. While president, Jefferson sponsored the Lewis and Clark expedition in 1803 to advance the geographic knowledge of the Nation, thus making clear that "the promotion of the general welfare depended heavily upon advances in scientific knowledge" (Unesco 1968, p. 11). In fact, this action by Jefferson set several important precedents including the provision of federal funds to individuals for scientific endeavors.

Although the Constitution did not set forth mechanisms for establishing national academic institutions, based on the founders' belief that the government should have only an indirect influence on science and technical advancement, the need for a national institution related to science and technology was recognized soon after the Revolutionary War. For example, West Point was founded in 1802 as the first national institution of a scientific and technical nature, although Connecticut established the first State Academy of Arts and Sciences in 1799.

In the early 1800s, universities began to emphasize science and technical studies, and in 1824 Rensselaer Polytechnic Institute was founded in New York State to emphasize the application of science and technology.

The *American Journal of Science* was the first American scientific publication, followed in 1826 by the *American Mechanics Magazine*.

The social importance of the government having a direct role in the creation and application of technical knowledge was emphatically demonstrated in the 1820s and 1830s through its support of efforts to control the cholera epidemic of 1822. Also during that time period, federal initiatives were directed toward manufacturing and transportation. In fact, the Secretary of the Treasury—the Department of the Treasury being the most structured executive department at that time—directly funded the Franklin Institute in Philadelphia to investigate the causes of these problems. This action, driven by public concern as well as the need to develop new technical knowledge, was the first instance of the government sponsoring research in a private-sector organization.

In 1838, the federal government again took a lead in the sponsorship of a technological innovation that had public benefits. After Samuel Morse demonstrated the feasibility of the electric telegraph, Congress provided him with $30,000 to build an experimental line between Baltimore, Maryland, and Washington, DC. This venture was the first instance of governmental support to a private researcher.

In retrospect, one could make an argument that Jefferson's funding of Lewis and Clark was the first instance of public support for pure research, whereas Morse was funded to conduct applied research. Although not discussed herein, there are other historical examples of governmental support to individuals for research that had the potential to benefit society, such as the Longitude Act of 1714. The British Parliament offered a prize (equal to several million dollars in today's terms) for a practicable solution for sailing vessels to determine longitude (Sobel 1995).

Public/private research relationships continued to evolve in frequency and in scope. In 1829, James Smithson, gifted $500,000 to the United States to found an institution in Washington, DC for the purpose of "increasing and diffusing knowledge among men" (Unesco, p. 12). Using the Smithson gift as seed funding, Congress chartered the Smithsonian Institution in 1846, and Joseph Henry became its first Executive Officer. Henry, a renowned experimental physicist, continued the practice of a federal agency directly supporting research through grants to individual investigators to pursue fundamental research. Also, the Institution represented a base for external support of scientific and engineering research; during the 1850s, about 100 academic institutions were established with science and engineering emphases.

Thus, the pendulum had made one complete swing in the hundred or so years since the signing of the Constitution. In the early years, the government viewed itself as having no more than an indirect influence on the development of science and technology, but over time its role changed from indirect to direct. This change was justified in large part because advances in science and technology came to be viewed as critical in promoting the public interest. This changing pattern of advocacy during the colonial period is summarized in Table 2.1.

THE PERIOD OF NATIONAL SCIENCE AND TECHNOLOGY INFRASTRUCTURE

Scientists had long looked toward the European universities for training in the sciences, but in the early and mid-1800s an academic infrastructure was beginning to develop in the United States. Harvard University awarded its first bachelors of science degree in 1850.

The development of an academic science base and the birth of technology-based industries (e.g., the electrical industry) in the late 1850s established what would become the foundation for America's technological preeminence.

Table 2.1. Pendulum Swing of Government's Role during the Colonial Period

Direct Role for the Government	Indirect Role for the Government
	1787 Constitutional Convention: establishment of national university for promotion of science rejected in favor of an indirect influence
	1790 Based on Article I, Section 8 of the Constitution, Congress passed the Patent Act
1803 President Jefferson commissioned the Lewis and Clark expedition	
1822 National initiatives in response to medical emergencies	
1824 ff. States began to establish science and technology universities	
1838 Direct funding to Samuel Morse to build a telegraph line between Washington, DC and Baltimore, MD	

The Morrill Act of 1862 established the land grant college system thereby formally recognizing the importance of trained individuals in the agricultural sciences. The Act charged each state to establish at least one college in the agricultural and mechanical sciences. Each state was given 30,000 acres of federal land per each elected U.S. Senator and

Representative. An important outgrowth of this land grant system was a mechanism or infrastructure through which state and federal governments could financially support academic research interests.

In 1863, during the Civil War, Congress established the National Academy of Sciences. The federal government funded the Academy but not the members affiliated with it who had (Unesco 1968, p. 14):

> ... an obligation to investigate, examine, experiment, and report upon any subject of science or art in response to a request from any department of the Government.

Then, as today, the Academy was independent of governmental control.

Although the government was encouraging an infrastructure to support science and technical research, it did not have a so-called in-house staff of permanent professionals who were competent to identify either areas of national importance or areas of importance to specific agencies. In 1884, Congress established the Allison Commission to consider this specific issue. While many solutions were debated, including the establishment of a Department of Science—an idea that resurfaces every few decades—the Commission soon disbanded without making any recommendations much less reaching closure on the matter. One could conclude from the inaction of the Commission that it favored the decentralized administrative architecture that had evolved over time as opposed to a centralized one.

The changing pattern of advocacy during the period of infrastructural growth is summarized in Table 2.2.

THE PERIOD OF INDUSTRIAL SCIENCE AND TECHNOLOGY INFRASTRUCTURE

Most scientists in the United States in the 1870s and 1880s had been trained in Europe, Germany in particular. What they experienced firsthand were the strong ties between European industries and graduate institutions of learning. Companies invested in professors and in their graduate students by providing them with funding and access to expensive materials and instruments, and in return the companies gained lead-time toward new discoveries as well as early access to the brightest graduate students as soon as they completed their studies. This form of symbiotic arrangement became the norm for the European-trained scientists who were working in U.S. industries and U.S. universities toward the end of the century.

By the turn of the century, it was widely accepted among industrial leaders that scientific knowledge was the basis for engineering development and it was the key to remaining competitive. Accordingly, industrial research laboratories soon began to blossom as companies realized their need to foster scientific knowledge outside of the university setting. There are a number of examples of this strategy.[2]

Table 2.2. Pendulum Swing of Government's Role during the Infrastructural Growth Period

Direct Role for the Government	Indirect Role for the Government
1862 Morrill Act—established the land grant college system	
1863 National Academy of Sciences established	
	1884 Allison Commission did not recommend the establishment of a federal Department of Science

General Electric (GE) established the General Electric Research Laboratory in 1900 in response to competitive fears that improved gas lighting would adversely affect the electric light business, and that other electric companies would threaten GE's market share as soon as the Edison patents expired.

Similarly, AT&T was at the same time facing increasing competition from radio technology. In response, AT&T established Bell Laboratories to research new technology in the event that wire communications were ever challenged.

And as a final example, Kodak realized at the turn of the century that it must diversify from synthetic dyes. For a number of years Kodak relied on German chemical technology, but when that technology began to spill

---

[2] Hounshell (1996) provides an excellent history of the growth of U.S. industrial research organizations.

over into other areas, such as photographic chemicals and film, Kodak realized that their competitive long-term health rested on their staying ahead of their rivals. Kodak, too, formed an in-house research laboratory.

Many smaller firms also realized the competitive threats that they could potentially face as a result of technological competition, but because of their size they could not afford an in-house facility. So as a market response, contract research laboratories began to form. Arthur D. Little was one such contract research laboratory that specialized in the area of chemicals.

Just as industrial laboratories were growing and being perceived by those in both the public and private sectors as vitally important to the economic health of the Nation, private foundations also began to grow and to support university researchers. For example, the Carnegie Institution of Washington was established in 1902, the Russell Sage Foundation in 1907, and the Rockefeller Foundation in 1913.

In the early-1900s science and technology began to be embraced—both in concept and in practice—by the private sector as the foundation for long-term competitive survival and general economic growth.

THE PERIOD OF THE WORLD WARS AND AFTERWARDS

Increased pressure on the pace of scientific and technical advancements came at the beginning of World War I. The United States had been cut off from its European research base. Congress, in response, established the Council of National Defense in 1916 to identify domestic pockets of scientific and technical excellence.

The National Academy of Sciences recommended to President Woodrow Wilson the formation of the National Research Council to coordinate cooperation between the government, industry, and the academic communities toward common national goals. The Allison Commission had failed in 1884 to formulate an infrastructure to undertake this task.

The prosperity of the post-World War I decade also created an atmosphere supportive of the continued support of science and technology. In 1920, there were about 300 industrial research laboratories, and by 1930 there were more than 1,600. Of the estimated 46,000 practicing scientists in 1930, about half were at universities and over a third were in industry. Herbert Hoover was Secretary of Commerce at this time. He adopted the philosophy that (Unesco 1968, p. 18):

... pure and applied scientific research constitute a foundation and instrument for the creation of growth and efficiency of the economy.

Two important events occurred in 1933 in response to the Great Depression and the subsequent national economic crisis. One event was the appointment of a Science Advisory Board, and the other event was the establishment of a National Planning Board. Whereas the National Research Council had been organized around *fields of science* to address governmental needs, the Science Advisory Board was multi-field and organized around *impending national problems*. The National Planning Board was formed on the presumption that there were areas of economic concern that required a national perspective rather than a field-of-science perspective.

In 1934, the National Resources Committee replaced the National Planning Board, and the Committee then subsumed the Science Advisory Board. The bottom line, after all of the organizational issues were settled, was that the federal government recognized through the formation of these committees and boards that it had and would continue to have an important coordinating role to play in science and technology planning toward a national goal of economic well being. Hence, the pendulum began to swing again, this time away from government having a hands-on role toward it having an indirect influence on planning the environment for science and technology.

In 1938, the Science Committee of the National Resources Committee issued a multi-volume report entitled, *Research—A National Resource*. Some important first principles were articulated in that report. Since then, these principles have formed a basis for economists and policy makers to rationalize and justify, again, a direct role of government in science and technology. The report is explicit that:

- There are certain fields of science and technology which the government has a Constitutional responsibility to support. These fields include defense, determination of standards, and certain regulatory functions.
- The government is better equipped to perform research in certain fields of science than the private sector. These are areas where "research is unusually costly in proportion to its monetary return but is of high practical or social value" (p. 25). Examples cited in the report include aeronautical and geological research.

- Research by the government "serves to stimulate and to catalyze scientific activity by nongovernmental agencies. In many fields, new lines of research are expensive and returns may be small or long delayed. Industry cannot afford to enter such fields unless there is reasonable prospect of definite financial gain within a predictable future, and it is under such circumstances that the Government may lead the way." (p. 26). One example cited was the Navy Department's influence on the development of the steel industry.

The involvement of the United States in World War II had a dramatic impact on the scope and direction of government's support of science and technology. Prior to the war, there were about 92,000 scientists, with about 20 percent in government and the remaining 80 percent being almost equally divided between universities and the more than 2,200 industrial laboratories. Clearly, the United States had a significant scientific resource base to draw upon for its war efforts.

In 1940, President Franklin D. Roosevelt established the National Defense Research Committee, and he asked Vannevar Bush, President of Carnegie Institution of Washington, to be its chairman. The purpose of this committee was to organize scientific and technological resources toward enhancing national defense. It soon became apparent that this task required an alternative administrative structure.

In 1941, Roosevelt issued an Executive Order establishing the Office of Scientific Research and Development (OSRD) with Bush as Director. The OSRD did not conduct research, rather it realized that there were pockets of scientific and technological excellence throughout the country, and through contractual relationships with universities and industry and government agencies, it could harness national strengths with a focus on ending the war. One hallmark event from the efforts of the OSRD was the establishment of the Los Alamos Laboratory in New Mexico under the management of the University of California. What came about from the collective efforts of the resources acquired by the Office were not only atomic weapons but also radar.

It was clear by 1944 that World War II was almost over. President Roosevelt then asked Bush to develop recommendations as to how scientific advancements could contribute in the larger sense to the advancement of national welfare. In his November 17, 1944 letter to Bush, President Roosevelt stated:

> The Office of Scientific Research and Development, of which you are the Director, represents a unique

experiment of team-work and cooperation in coordinating scientific research and in applying existing scientific knowledge to the solution of the technical problems paramount in war. ... There is ... no reason why the lessons to be found in this experiment cannot be profitably employed in times of peace. This information, the techniques, and the research experience developed by the Office of Scientific Research and Development and by the thousands of scientists in the universities and in private industry, should be used in the days of peace ahead for the improvement of the national health, the creation of new enterprises bringing new jobs, and the betterment of the national standard of living. ... New frontiers of the mind are before us, and if they are pioneered with the same vision, boldness, and drive with which we have waged this war we can create a fuller and more fruitful employment and a fuller and more fruitful life.

Shortly before asking Bush to prepare this report, Senator Harley M. Kilgore from West Virginia had introduced a bill to create a National Science Foundation. The Kilgore bill recommended giving authority to federal laboratories to allocate public moneys in support of science to other government agencies and to universities. Clearly, this recommendation gave a direct role to government in shaping the technological course of the country not only in terms of scientific direction but also in terms of what groups would conduct the underlying research. The bill was postponed until after the war.

Bush submitted his report, *Science—the Endless Frontier*, to President Roosevelt on July 25, 1945. In Bush's transmittal letter to the president he stated:

The pioneer spirit is still vigorous within this Nation. Science offers a largely unexplored hinterland for the pioneer who has the tools for his task. The reward of such exploration both for the Nation and the individual are great. Scientific progress is one essential key to our security as a nation, to our better health, to more jobs, to a higher standard of living, and to our cultural progress.

The foundations set forth in *Science—the Endless Frontier* are:

- "Progress ... depends upon a flow of new scientific knowledge" (p. 5).
- "Basic research leads to new knowledge.[3]   It provides scientific capital. ... New products and new processes do not appear full-grown. They are founded on new principles and new conceptions, which in turn are painstakingly developed by research in the purest realms of science" (p. 11).
- "The responsibility for the creation of new scientific knowledge ... rests on that small body of men and women who understand the fundamental laws of nature and are skilled in the techniques of scientific research" (p. 7).
- "A nation which depends upon others for its new basic scientific knowledge will be slow in its industrial progress and weak in its competitive position in world trade, regardless of its mechanical skill" (p. 15).
- "The Government should accept new responsibilities for promoting the flow of new scientific knowledge and the development of scientific talent in our youth" (p. 7).
- "If the colleges, universities, and research institutes are to meet the rapidly increasing demands of industry and Government for new scientific knowledge, their basic research should be strengthened by use of public funds" (p. 16).
- "Therefore I recommend that a new agency for these purposes be established" (p. 8).

Bush recommended in his report the creation of a National Research Foundation. Its proposed purposes were to:

> ... develop and promote a national policy for scientific research and scientific education, ... support basic research in nonprofit organizations, ... develop scientific talent in American youth by means of scholarships and fellowships, and ... contract and otherwise support long-range research on military matters.

Bush envisioned a National Research Foundation that would provide funds to institutions outside government for the conduct of research. Thus, this organization differed from Kilgore's proposed National Science

---

[3] The term "basic research" is credited to Vannevar Bush. He proffered the definition: "Basic research is performed without thought of practical ends."

Foundation in that Bush advocated an indirect role for government. There was agreement throughout government that an institutional framework for science was needed, but the nature and emphases of that framework would be debated for yet another five years.

Science—the Endless Frontier affected the scientific and technological enterprise of this Nation in at least two ways. It laid the basis for what was to become the National Science Foundation in 1950. Also, it set forth a paradigm that would over time influence the way that policy makers and academic researchers thought about the process of creating new technology. The so-called linear model set forth by Bush is often represented by:

Basic Research → Applied Research → Development → Enhanced Production → Economic Growth

Complementing *Science—the Endless Frontier* was a second, and often overlooked, report prepared in 1947 by John Steelman, then Chairman of the President's Scientific Research Board. As directed by an Executive Order from President Harry Truman, Steelman, in *Science and Public Policy*, made recommendations on what the federal government could do to meet the challenge of science and assure the maximum benefits to the Nation. Steelman recommended that national R&D expenditures should increase as rapidly as possible, citing (p. 13):

1. Need for Basic Research.
   Much of the world is in chaos. We can no longer rely as we once did upon the basic discoveries of Europe. At the same time, our stockpile of unexploited fundamental knowledge is virtually exhausted in crucial areas.
2. Prosperity.
   This Nation is committed to a policy of maintaining full employment and full production. Most of our frontiers have disappeared and our economy can expand only with more intensive development of our present resources. Such expansion is unattainable without a stimulated and growing research and development program.
3. International Progress.
   The economic health of the world—and the political health of the world—are both intimately associated

with our own economic health. By strengthening our economy through research and development we increase the chances for international economic well-being.

4. Increasing Cost of Discovery.
    The frontiers of scientific knowledge have been swept so far back that the mere continuation of pre-war growth, even in stable dollars, could not possibly permit adequate exploration. This requires more time, more men, more equipment than ever before in industry.

5. National Security.
    The unsettled international situation requires that our military research and development expenditures be maintained at a high level for the immediate future. Such expenditures may be expected to decrease in time, but they will have to remain large for several years, at least.

An important element of the Steelman report was the recommended creation of a National Science Foundation, similar in focus to the National Research Foundation outlined by Bush. And, Congress passed the National Science Foundation Act in 1950.

Renewed post-war attention toward science and technology came with the success of the Soviet Union's space program and the orbit of its Sputnik I in October 1957. In response, President Dwight D. Eisenhower championed a number of committees and agencies to ensure that the United States could soon be at the forefront of this new frontier. Noteworthy was the National Defense Education Act of 1958, which authorized $1 billion in federal moneys for support of science, mathematics, and technology graduate education. This proposal is precisely the type of support that Bush recommended in his report.

As the post-World War II period came to close, there was a well-established national and industrial infrastructure to support the advancement of science and technology. But, more important than the infrastructure, there was an imbedded belief that scientific and technological advancements are fundamental for economic growth, and that the government has an important supporting role—both direct and indirect—to ensure such growth.

The changing pattern of advocacy during the period of the World Wars, and afterwards, is summarized in Table 2.3.

Table 2.3. Pendulum Swing of Government's Role during the World Wars Period and Afterwards

Direct Role for the Government	Indirect Role for the Government
1938 National Resources Committee report, *Research—A National Resource*	
	1945 Vannevar Bush's report, *Science—the Endless Frontier*
	1950 National Science Foundation established

Every president since President Eisenhower has initiated at least one major science and technology policy initiative. Representative initiatives are:

- President John F. Kennedy set the goal of sending a man to the moon by the end of the 1960s and funded the needed programs to make this a reality.
- President Lyndon B. Johnson emphasized the use of scientific knowledge to solve social problems through, for example, his War on Poverty.
- President Richard M. Nixon dramatically increased federal funding for biomedical research as part of his War on Cancer.
- President Gerald R. Ford created the Office of Science and Technology Policy (OSTP) within the Executive Branch.
- President James E. Carter initiated research programs for renewable energy sources such as solar energy and fission.
- During President Ronald W. Reagan's administration, expenditures on defense R&D increased dramatically as part of his Star Wars system.
- President George H. W. Bush (not related to Vannevar Bush) set forth this Nation's first technology policy and increased the scope of the National Institute of Standards and Technology (NIST).
- President William J. Clinton established important links between science and technology policy, championing programs to transfer publicly-funded technology to the private sector.

- President George W. Bush advocated making the R&E tax credit permanent.

3 PUBLIC SUPPORT OF INNOVATION

The government has historically had, as briefly overviewed in Chapter 2, and should continue to have an important partnership role with the private sector in fostering innovation. This intuitive conclusion logically follows from these facts:

- Innovation leads to technology.
- Technology is the prime driver of economic growth.
- In the absence of government intervention, firms will underinvest in the innovation process, especially in R&D.
- Government has a responsibility to address this underinvestment by providing incentives for the continued conduct of, or perhaps increase in, R&D.

Such sequential reasoning to justify the role of government in innovation has dominated the history of public-sector involvement in the innovation process, and more recently of the growth of public/private partnerships as related to innovation. And, the focus of R&D in this sequence of thought reflects upon the linear model in Chapter 2 wherein R&D leads to enhanced production and enhanced production leads to economic growth:

$$R\&D \rightarrow Enhanced\ Production \rightarrow Economic\ Growth$$

However, the economic underpinnings of government's role in innovation are more complex than the above logic might suggest.[1]

---

[1] This chapter draws directly from Link and Scott (2004, forthcoming a).

GOVERNMENT'S ROLE IN INNOVATION

The theoretical basis for government's role in market activity is based on the concept of market failure. Market failure is typically attributed to market power, imperfect information, externalities, and public goods. The explicit application of market failure to justify government's role in innovation—in R&D activity in particular—is a relatively recent phenomenon within public policy.

Many point to President George H.W. Bush's 1990 *U.S. Technology Policy* as the Nation's first formal domestic technology policy statement. Albeit an important initial policy effort, it however failed to articulate a foundation for government's role in innovation and technology. Rather, it implicitly assumed that government had a role, and then set forth the general statement (1990, p. 2):

> The goal of U.S. technology policy is to make the best use of technology in achieving the national goals of improved quality of life for all Americans, continued economic growth, and national security.

President William Clinton took a major step forward from the 1990 policy statement in his 1994 *Economic Report of the President* by articulating first principles about why government should be involved in the technological process (1994, p. 191):

> The goal of technology policy is not to substitute the government's judgment for that of private industry in deciding which potential 'winners' to back. Rather, the point is to correct market failure ...[2]

Subsequent Executive Office policy statements have echoed this theme; *Science in the National Interest* (1994) and *Science and Technology: Shaping the Twenty-First Century* (1998) are among such examples. President Clinton's 2000 *Economic Report of the President* (2000, p. 99) elaborated upon the concept of market failure as part of U.S. technology policy:

---

[2] The conceptual importance of identifying market failure for policy is also emphasized, although without any operational guidance, in Office of Management and Budget (1996).

Rather than support technologies that have clear and immediate commercial potential (which would likely be developed by the private sector without government support), government should seek out new technologies that will create benefits with large spillovers to society at large.

Relatedly, Martin and Scott (2000, p. 438) observed:

Limited appropriability, financial market failure, external benefits to the production of knowledge, and other factors suggest that strict reliance on a market system will result in underinvestment in innovation, relative to the socially desirable level. This creates a *prima facie* case in favor of public intervention to promote innovative activity.

Underinvestment in R&D

Market failure, as addressed in this chapter, and of the type which could specifically be termed technological or innovation market failure, refers to a condition under which the market, including both the R&D-investing producers of a technology and the users of the technology, underinvests, from society's standpoint, in a particular technology. Such underinvestment occurs because conditions exist that prevent organizations from fully realizing or appropriating the benefits created by their investments.

The following explanation of market failure and the reasons for market failure follow closely Arrow's (1962) seminal work in which he identified three sources of market failure related to knowledge-based innovative activity—"indivisibilities, inappropriability, and uncertainty" (p. 609).[3]

To explain, consider a marketable technology to be produced through an R&D process where conditions prevent full appropriation of the benefits from technological advancement by the R&D-investing firm. Other firms in the market or in related markets will realize some of the

---

[3] Although Arrow does not elaborate on indivisibilities and inappropriability in his paper, the concepts are well understood in the innovation literature. Recalling that Arrow defines innovation "as the production of knowledge" (1962, p. 609), the market does not price knowledge in discrete bundles and thus because of such indivisibilities market prices may not send appropriate signals for economic units to make marginal decisions correctly.

profits from the innovation, and of course consumers will typically place a higher value on a product than the price paid for it. The R&D-investing firm will then calculate, because of such conditions, that the marginal benefits it can receive from a unit investment in such R&D will be less than could be earned in the absence of the conditions reducing the appropriated benefits of R&D below their potential, namely the full social benefits. Thus, the R&D-investing firm may underinvest in R&D, relative to what it would have chosen as its investment in the absence of the conditions. Stated alternatively, the R&D-investing firm may determine that its private rate of return is less than its private hurdle rate and therefore it will not undertake socially valuable R&D.

The basic concept can be illustrated with Figure 3.1, which follows from Tassey (1992, 1997, 1999) and Jaffe (1998). The social rate of return is measured on the vertical axis along with society's hurdle rate on investments in R&D. The private rate of return is measured on the horizontal axis along with the private hurdle rate on R&D. A 45-degree line (dashed line) is imposed on the figure under the assumption that the social rate of return from an R&D investment will at least equal the private rate of return from the same investment. Three separate R&D projects are labeled as project A, B, and project C. Each is shown, for illustrative purposes only, to have the same social rate of return.

For project A, the private rate of return is less than the private hurdle rate because of barriers to innovation and technology. As such, the private firm will not choose to invest in project A, although the social benefits from undertaking project A would be substantial. The same is true for project B although the private rate of return is closer to the private hurdle rate than for project A.

The principle of market failure illustrated in the figure relates to appropriability of returns to investment. The vertical distance shown with the double arrow for project A is called the spillover gap; it results from the additional value society would receive above what the private firm would receive if project A were undertaken. What the firm would receive (along the 45-degree line) is less than its hurdle rate because the firm is unable to appropriate all of the returns that spill over to society. Project A is the type of project in which public resources should be invested to ensure that the project is undertaken. The level of public resources necessary to ensure that project B is undertaken would be less than for project A.

In comparison, project C yields the same social rate of return as projects A and B, but most of that return can be appropriated by the innovator, and the private rate of return is greater than the private hurdle rate. Hence, project C is one for which the private sector has an incentive to invest on its own or, alternatively stated, there is no economic justification for public resources being allocated to support project C.

Figure 3.1. Spillover Gap between Social and Private Rates of Return to R&D

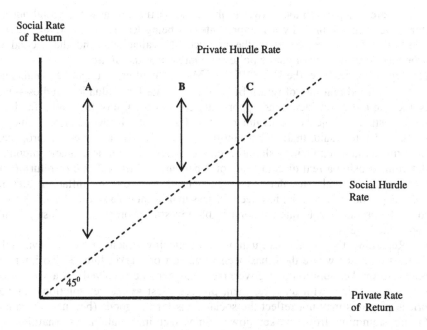

For projects of type A where significant spillovers occur, government's role has typically been to provide funding or technology infrastructure through public research institutions that lowers the marginal cost of investment so that the marginal private rate of return exceeds the private hurdle rate. Since the private return to project B is closer to the private hurdle rate, incentives might be an appropriate policy tool.

Note that the private hurdle rate is greater than the social hurdle rate in the figure. This is primarily because of management's (and employees') risk aversion and issues related to the availability and cost of capital. These factors represent an additional source of market failure that is related to uncertainty. For example, because most private firms are risk averse (i.e., the penalty from lower than expected returns is weighted more heavily than the benefits from greater than expected returns), they require

a higher hurdle rate of return compared to society as a whole that is closer
to being risk neutral.[4]

To reduce market failures associated with inappropriability and
uncertainty, government typically engages in activities to reduce technical
and market risk (actual and perceived). These activities include, but are
not limited to, the activities of public research institutions, as discussed

---

[4] There are two parts to the answer to the twin questions of how the social hurdle
rate is determined and why it is represented as being less than the private hurdle
rate. The first is grounded in the practice of evaluations, and the second is
grounded in the theory of public policy to address market failure.

(1) Regarding practice, the U.S. Office of Management and Budget has mandated
that a specified real rate of return be used as the rate for evaluation studies—that
is, the rate to be considered the opportunity cost for the use of the public funds in
the investment projects we evaluate. The Office of Management and Budget
(1992, p.9) has said that: "Constant-dollar benefit-cost analyses of proposed
investments and regulations should report net present value and other outcomes
determined using a real discount rate of 7 percent." That real rate of return (and
the related nominal rates derived by accounting for expected inflation rates in
various periods of analysis) has been far less than what case studies have reported
to be the private hurdle rate for comparable investment projects in industry (Link
and Scott 1998b).

(2) Regarding theory, the evaluation of public investment projects, invariably
focuses on cases where there has been some sort of market failure. To improve
upon the market solution, the government has become involved (in a variety of
ways, in practice) with an investment project. Just as market solutions for the
prices of goods may not reflect the social costs for the goods (because of market
failure stemming from market power, imperfect information, externalities, or
public goods), the private hurdle rates that reflect market solutions for the price of
funds—the opportunity cost of funds to the private firms—may not reflect the
social cost of the funds. The government may decide that the appropriate social
cost—the opportunity cost for the public funds to be invested—differs from the
market solution. Typically, in practice, the government believes that it faces less
risk than the private sector firms doing similar investments; hence it will believe a
lower yield is satisfactory since the public is bearing less risk than the private
sector firm going it alone with a similar investment. More generally, government
must decide what the opportunity costs of its public funds will be in various uses,
and in general that will not be the same as the market rate. However, all that said,
it is known from Arrow's thinking about social choice that the government's
decision about what the rate should be cannot possibly reflect the diversity of
opinion in the private sector regarding the decision (Arrow 1963). Consequently,
as a logical matter, one could not prove that the government's choice of the right
hurdle rate is obviously correct because diversity of opinion about the correct rate
will not be reflected in the government's choice.

below. The following section discusses several circumstances—termed barriers to technology—that cause market failure and an underinvestment in R&D.

Barriers to Innovation and Technology

There are a number of factors that can explain why a firm will perceive that its expected private rate of return will fall below its hurdle rate.[5] Individuals will differ not only about a listing of such factors because they are not generally mutually exclusive, but also they will differ about the relative importance of one factor compared to another in whatever taxonomy is chosen.

First, high technical risk (that is, outcomes may not be technically sufficient to meet needs) may cause market failure given that when the firm is successful, the private returns fall short of the social returns. The risk of the activity being undertaken is greater than the firm can accept, although if successful there would be very large benefits to society as a whole. Society would like the investment to be made, but from the perspective of the firm, the present value of expected returns is less than the investment cost and is thus less than the amount yielding its acceptable return on investment.

Second, high technical risk can relate to high commercial or market risk (although technically sufficient, the market may not accept the innovation—reasons can include factors listed subsequently such as imitation or competing substitutes or interoperability issues) as well as to technical risk when the requisite R&D is highly capital intensive. The project may require too much capital for any one firm to feel comfortable with the outlay. The minimum cost of conducting research is thus viewed as excessive relative to the firm's overall R&D budget, which considers

---

[5] As Arrow (1962) explained, investments in knowledge entail uncertainty of two types—technical and market. The technical and market results from technology may be very poor, or perhaps considerably better than the expected outcome. Thus, a firm is justifiably concerned about the risk that its R&D investment will fail, technically or for any other reason. Or, if technically successful, the R&D investment output may not pass the market test for profitability. Further, the firm's private expected return typically falls short of the expected social return as previously discussed. This concept of downside risk is elaborated upon in Link and Scott (2001).

the costs of outside financing and the risks of bankruptcy. In this case, the firm will not make the investment, although society would be better off if it had, because the project does not appear to be profitable from the firm's private perspective.

Third, many R&D projects are characterized by a lengthy time interval until a commercial product reaches the market. The time expected to complete the R&D and the time until commercialization of the R&D results are long, and the realization of a cash flow from the R&D investment is in the distant future. If a private firm faces greater risk than society does, and as a result requires a greater rate of return and hence applies a higher discount rate than society does, it will value future returns less than does society. Because the private discount rate exceeds the social discount rate, there may be underinvestment, and the underinvestment increases as the time to market increases because the difference in the rate is compounded and has a bigger effect on returns further into the future.

Fourth, it is not uncommon for the scope of potential markets to be broader than the scope of the individual firm's market strategies so the firm will not perceive or project economic benefits from all potential market applications of the technology. As such, the firm will consider in its investment decisions only those returns that it can appropriate within the boundaries of its market strategies. While the firm may recognize that there are spillover benefits to other markets, and while it could possibly appropriate them, such benefits are ignored or discounted heavily relative to the discount weight that would apply to society.

A similar situation arises when the requirements for conducting R&D demand multidisciplinary research teams; unique research facilities not generally available with individual companies; or fusing technologies from what were heretofore separate, non-interacting parties. The possibility for opportunistic behavior in such thin markets may make it impossible, at a reasonable cost, for a single firm to share capital assets even if there were not R&D information sharing difficulties to compound the problem. If society, perhaps through a technology-based public institution, could act as an honest broker to coordinate a cooperative multi-firm effort, then the social costs of the multidisciplinary research might be less than the market costs.[6]

Fifth, the evolving nature of markets requires investments in combinations of technologies that, if they existed, would reside in different industries that are not integrated. Because such conditions often transcend

<hr>

[6] See Leyden and Link (1999) on the role of a federal laboratory as an honest broker in the innovation process.

the R&D strategy of firms, such investments are not likely to be pursued. That is not only because of the lack of recognition of possible benefit areas or the perceived inability to appropriate whatever results, but also because coordinating multiple players in a timely and efficient manner is cumbersome and costly. Again, as with the multidisciplinary research teams, society may be able to use a technology-based public institution to act as an honest broker and reduce costs below those that the market would face.

Sixth, a situation could exist when the nature of the technology is such that it is difficult to assign intellectual property rights. Knowledge and ideas developed by a firm that invests in technology may spill over to other firms during the R&D phase or after the new technology is introduced into the market. If the information creates value for the firms that benefit from the spillovers, then other things being equal, the innovating firms may underinvest in the technology.

Relatedly, when competition in the development of new technology is very intense, each firm, knowing that the probability of being the successful innovator is low, may not anticipate sufficient returns to cover costs. Further, even if the firm innovates, intense competition at the application stage can result because of competing substitute goods, whether patented or not. Especially when the cost of imitation is low, an individual firm will anticipate such competition and may therefore not anticipate returns sufficient to cover its R&D investment costs.

Of course, difficulties appropriating returns need not always inhibit R&D investment (Baldwin and Scott 1987). First-mover advantages associated with customer acceptance and demand, as well as increasing returns as markets are penetrated and production expanded, can imply that an innovator wins most (or at least a sufficient portion to support the investment) of the rewards even if it does not "take all."

Seventh, industry structure may raise the cost of market entry for applications of the technology. The broader market environment in which a new technology will be sold can significantly reduce incentives to invest in its development and commercialization because of what some scholars have called technological lock-in and path dependency.[7] Many technology-based products are part of larger systems of products. Under such industry structures, if a firm is contemplating investing in the development of a new product but perceives a risk that the product, even if

---

[7] See David (1987) for detailed development of the ideas of path dependency in the context of business strategies and public policy toward innovation and diffusion of new technologies.

technically successful, will not interface with other products in the system, the additional cost of attaining compatibility or interoperability may reduce the expected rate of return to the point that the project is not undertaken.

Similarly, multiple sub-markets may evolve, each with its own interface requirements, thereby preventing economies of scale or network externalities from being realized. Again, society, perhaps through a technology-based public institution, may be able to help the market's participants coordinate successful compatibility and interoperability.

Eighth, situations exist where the complexity of a technology makes agreement with respect to product performance between buyer and seller costly. Sharing of the information needed for the exchange and development of technology can render the needed transactions between independent firms in the market prohibitively costly if the incentives for opportunistic behavior are to be reduced to a reasonable level with what Teece (1980) called obligational contracts.

Teece emphasized that the successful transfer of technology from one firm to another often requires careful teamwork with purposeful interactions between the seller and the buyer of the technology. In such circumstances, both the seller of the technology and the buyer of the technology are exposed to hazards of opportunism. Sellers, for example, may fear that buyers will capture the know-how too cheaply or use it in unexpected ways. Buyers may worry that the sellers will fail to provide the necessary support to make the technology work in the new environment; or they may worry that after learning about the buyer's operations in sufficient detail to transfer the technology successfully, the seller would back away from the transfer and instead enter the buyer's industry as a technologically sophisticated competitor.

Once again, if society can use a technology-based public institution to act as an honest broker, the social costs of sharing technology may be less than market costs.

These eight factors that create, individually or in combination, barriers to innovation and technology and thus lead to a private underinvestment in R&D are listed in Table 3.1. While these factors have been discussed individually above, and are listed in the table as if they are discrete phenomena, they are interrelated and overlapping, although in principle any one factor could be sufficient to cause a private firm to underinvest in R&D.

THE ROLE OF PUBLIC RESEARCH INSTITUTIONS

Public research institutions—their intramural research as well as their focused extramural research activity—could overcome many of the barriers to innovation and technology discussed in the previous section.

Table 3.1. Factors Creating Barriers to Innovation and Technology

1.	High technical risk associated with the underlying R&D
2.	High capital costs to undertake the underlying R&D
3.	Long-time to complete the R&D and commercialize the resulting technology
4.	Underlying R&D spills over to multiple markets and is not appropriable
5.	Market success of the technology depends on technologies in different industries
6.	Property rights cannot be assigned to the underlying R&D
7.	Resulting technology must be compatible and interoperable with other technologies
8.	High risk of opportunistic behavior when sharing information about the technology

For the purpose of describing the rationale for public research institutions that provide, intramurally or extramurally, infrastructure technology needed in public/private partnerships, a definition of risk is posited that is focused on the operational concern with the downside outcomes for an investment. The shortfalls of the private expected outcomes from society's expected returns reflect appropriability problems.

There are several related technological and market factors that will cause private firms to appropriate less return and to face greater risk than society does. These factors underlie what Arrow (1962) identified as the non-exclusivity and public good characteristics of investments in the creation of knowledge. The private firms' incomplete appropriation of social returns in the context of technical and market risk can make risk in its operational sense unacceptably large for the private firm considering an investment.

Operationally and with reference to Figure 3.1, Tassey (1992, 1997, 2005), for example, defines risk as the probability that a project's rate of return falls below a required, private rate of return or private hurdle rate (as opposed to simply deviating from an expected return).[8] As illustrated in Link and Scott (2001)—both in concept and in terms of the specific projects performed by the private sector with subsidies and oversight from the Advanced Technology Program (ATP) within the U.S. National Institute of Standards and Technology (NIST)—for many socially desirable investments, the private firm faces an unacceptably large probability of a rate of return that falls short of its private hurdle rate. Yet, from society's perspective, the probability of a rate of return that is less than the social hurdle rate is sufficiently small that the project is still worthwhile.

Martin and Scott (2000, pp. 438-439) make the point that the design of appropriate public policy should match the policy with the specific source of underinvestment. In that light, they identify several roles for public research institutions. Given the types of research they perform, such institutions could be called standards and infrastructure technology institutions. Specific activities of those institutions are matched with specific sources of underinvestment in research, and the various activities are illustrated with examples from case studies.

One role for a public research institution within a national innovation system is to facilitate the promulgation and adoption of standards and thereby, for example, reduce the risk associated with standards for new technology as inputs are developed for using industries such as in the sectors developing software, equipment, and instruments. The term standards is used in this context in a general sense to refer to voluntary performance protocols and interoperability standards, test methods, and standard reference materials. Although one can find examples where observers have thought that product standards were used in anticompetitive ways, the role for public research institutions is quite general and important, encompassing several types of standards. The public institution with research capability can respond to industry's needs for standards, working with industry to develop them while serving as an honest broker providing impartial mediation of disputes that could not be provided by a private firm with a proprietary interest in the outcomes.[9] In

---

[8] Tassey (1992, 1997, 2005) has developed and applied the idea of barriers to innovation and technology to a number of national policies.

[9] Industry's scientists and engineers frequently interact with scientists in the public research institution in conferences and workshops and together they enable the public research laboratories to develop the standards needed as the technological requirements for industry to remain competitive evolve. See the several examples described in Link and Scott (1998b, forthcoming a).

the absence of the public research institution, industry would have incurred higher costs to replace the public standards activities than the actual costs to the public institution for those activities. Further, the quality of the more costly private standards activities would have been less than the quality of the public standards activities.

For another role, public research institutions can oversee extension services to facilitate technology transfer in sectors such as light industry or agriculture when, for example, small firms, facing limited appropriability from their investments in new technologies yet providing large external benefits to the economy as a whole, apply inputs developed in supplying industries. Such extension services can make possible a vibrant entrepreneurial sector of smaller firms that stimulates the adoption and diffusion of new technology and also innovation, technological advance, and economic growth.

The positive impact of such an entrepreneurial sector has been documented by many scholars—for example, Audretsch (1995)—and in the last two decades recognition of its importance for economic growth has increased and become widespread. Imperfections in credit markets, opportunistic behavior by larger firms that might provide resources to small entrepreneurial businesses, and the unappropriated external benefits from entrepreneurial businesses might require public support of extension services to avoid underinvestment in the transfer of technologies. Although the argument for public research institutions with research capability is not as strong as it is in the foregoing role with standards, public research institutions such as NIST are in a good position to foster the technology transfer provided by extension programs. Such public institutions have knowledge of the key technologies, have working relationships with the industries supplying the technologies, and can assist with the transfer the technologies without opportunistic exploitation of the small firms, allowing them to grow as independent sources of initiative and growth.[10]

---

[10] Private organizations with some public funding have evidently been successful in transferring technology to smaller businesses. Although coordinated by a public research institution, there is substantial private funding for the Malcolm Baldrige National Quality Award Program through NIST. The Program is focused on improving management and competitiveness. The pattern of shared funding among government and private organizations is common to many of the activities of public research institutions—most prominently activities largely performed by the private sector with oversight from the public institution and with some partial public funding of the projects. See Link and Scott (forthcoming b).

For a third role, a public research institute can serve as the coordinator and facilitator for cooperative R&D efforts joining industry, universities, and government in research that is subsidized by the government. The several projects studied by Hall, Link, and Scott (2003) provide examples of such cooperative R&D efforts. Such cooperative research with a public research institution as the facilitator is often necessary to coordinate the development of infrastructure technologies as well as pre-competitive generic technologies that are at the heart of the development of complex systems involving high cost, risk, and limited appropriability. These complex systems are developed, for example, in aerospace, electrical and electronics technology, telecommunications, and computer technologies. While the coordination of cooperative efforts that transcend the solely market-based activities of industry is arguably an important and central role for government, the key question is whether a public research institution playing that coordinating role actually needs to have a research capacity itself. Based on case studies, it is clear that in many cases the answer is "Yes." For example, the ATP relies on the research capability of NIST not only to ensure sound oversight of the competitions for the government's chosen research projects that will be performed by industry with partial public funding, but also to provide coordinated research developing infrastructure technologies that support the advances in technology that the ATP hopes to foster through its awards for publicly subsidized and privately performed research.[11]

Finally, for industrial applications of technologies with high science content, where the knowledge base originates outside of the commercial sector, the creators of the knowledge may not recognize the potential applications or effectively communicate the new developments to potential users.

As a fourth role in a national innovation system, a public research institution can facilitate the diffusion of advances from research in these cases—such as in biotechnology, chemistry, materials science, and pharmaceuticals—where the applications have high science content. This fourth role is one of facilitating communication and dissemination of ideas from science that can then be used by many sectors to advance applied research and development. In many cases, government funds will have been used by universities to develop the basic science, because the ideas have a strong public good component and there would not have been sufficient incentive to develop them without government funding. Once the basic science is available, the knowledgeable public research

---

[11] See Link and Scott (2001, 2005a).

institution with expertise in both research and connections to industry can help to disseminate the information widely.

Granted that basic research with economy-wide implications and very long time horizons is unlikely to be undertaken by private firms, Are there reasons—incentive problems and market failures—that would require that the basic research should be *performed* by the government and not by, for example, the government's financing of private universities? That is, Are there reasons that the fourth role for the public research institution would include not only working to communicate basic science, transferring it to industry in ways that focus on the industrial usefulness of the basic science, but the public laboratory would actually do basic science itself. Although, experience with the work done in the U.S. government laboratories has revealed some fairly basic research, but even the most basic research is quite applied—using the basic science created in universities to develop new measurement technology for example. Conceivably there are incentive issues that may dictate the performance of certain types of basic research in the laboratories of public research institutions. By their nature, the research objectives of the government may differ from the interests of universities and their researchers, and it is possible that some goals of the government's basic research agenda would not align well with the current academic interests. Stated differently, academic researchers might find it beneath them to do the science that the public happens to want at a particular time. Another possibility is that academic researchers cannot always take the long view (especially given that the long view can change as political administrations change) needed to develop a government-mandated strand of science, in the detail needed, simply because of the constraints of turning out sufficient publications of sufficient variety and quality in the context of review and promotion for the researchers. Laboring in some public service vineyard for a decade or more may not have the necessary academic rewards to ensure survival within the university system. A public laboratory scientist is freed from such constraints and the public laboratory can set its own reward structure that is sensitive to the fact that political administrations change and the scientific imperatives of government can change. Finally, national security may dictate that some types of research are performed in government laboratories with heightened security rather than in the more open environment of university laboratories.

In all of the foregoing roles for the public research institution, the institution is not only an honest broker providing technological services—standards, standard reference materials, calibrations traceable to the standards, technology transfer and diffusion of scientific advances—without a proprietary, rivalrous, market-based interest. As well, the public research institution's research capability is an integral part of developing and maintaining standards and other technological services. The institution is not just an administrator, but an organization with real scientific and engineering expertise. In matters of generic and infrastructure technology, the institution is an honest broker with leading-edge research capabilities and close working relationships with industry allowing it to understand industry's needs and continually develop and maintain the standards and services that industry relies on for its productivity.

The theoretical foundation for public sector involvement in any aspect of the innovation process logically leads to a discussion of public accountability, meaning that the public sector is also responsible for evaluating the social benefits of its actions.

4 TECHNOLOGICAL CHANGE AND R&D

This chapter provides background on both production function-based models of technological change and R&D activity. Both topics are critical to an understanding of public/private partnerships.

MODELS OF TECHNOLOGICAL CHANGE

Much of the early literature on technological change stems from production function models in which the output (Q) of a microeconomic unit (a plant, a firm, or even an industry) is represented simply as a function of capital (K) and labor (L):

(1) $Q = f(K, L)$

Of course, there are other inputs in production, such as intermediate materials, services, and financial recourses, but for an initial exposition, a two input model is sufficient.

Following Hicks (1932), so-called Hicksian technological change is defined to be labor-saving, capital-saving, or neutral if the technological change brought about by the adoption of an underlying innovation raises, lowers, or leaves unchanged the marginal product of capital relative to the marginal product of labor for a given capital-to-labor ratio.[1]

---

[1] Harrod's (1948) classification scheme is similar to that of Hicks except that the capital-to-output ratio is assumed constant rather than the capital-to-labor ratio. Solow's (1967) classification is similar to Harrod's except that the labor-to-output ratio is assumed constant. See Link (1987) and Link and Siegel (2003) for a complete and mathematical overview of the production function concept of technological change.

Using the early production function models of technological change—or more accurately models of the classification of technological change because none of the models addressed the source of the innovation that brought about the technological change—Solow (1957) advanced the concept of an aggregate production function and illustrated it assuming that the function was Cobb-Douglas in nature:

(2) $Q = A(t) K^\alpha L^\beta$

where, assuming perfect competition and constant returns to scale, α and β ($\alpha + \beta = 1$) are the shares of income distributed to capital and labor respectively. From equation (2) it follows that the impact of technological change on production can be approximated as a residual growth rate measured as the percentage change in output less the percentage change in capital and labor. This Solow residual is often referred to as the percentage change total factor productivity (TFP), or simply productivity growth, and, based on equation (2), it is often denoted as $\acute{A}/A$.[2]

Since the early 1960s, researchers have engaged in empirical analyses to estimate the impact of investments in R&D on productivity growth under the implicit assumption that R&D is an input into innovation and innovation leads to technological change as measured by the growth in TFP.

For reference, TFP for the U.S. private non-farm business sector for the years 1948 through 2002 is shown in Figure 4.1. This figure will be referred to in later chapters because a number of public policy innovation initiatives were promulgated in response to slowdowns in TFP (in the early and mid-1970s and early 1980s).[3]

Conceptualizing the production function in equation (1) at the firm level, and introducing the firm's stock of technical capital, T, as a third input, the model becomes:

(3) $Q = A(t) F (K, L, T)$

---

[2] $\acute{A}$ denotes the time rate of change in TFP and thus $\acute{A}/A$ denotes the percentage rate of change in TFP.

[3] These data, as noted in Figure 4.1 come from the Bureau of Labor Statistics. The term the Bureau uses is multifactor productivity as opposed to total factor productivity.

If the source of the firm's technical capital is its R&D, then a model relating productivity growth to investments in R&D (*RD*) takes the form:[4]

(4) $\acute{A}/A = \lambda + \rho \, (RD/Q)$

Empirical estimates of ρ from equation (4) have been interpreted as an estimate of the marginal private rate of return to investments in R&D.[5]

Figure 4.1. U.S. Private Non-farm TFP Index, 1948-2002: (2000 = 100.0)

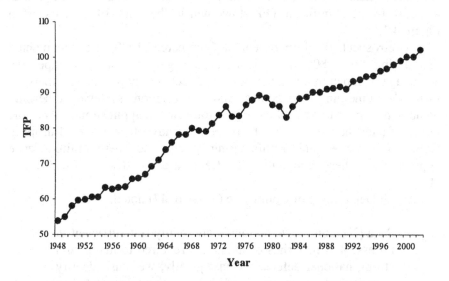

Source: Bureau of Labor Statistics.

As discussed in the next section, there are two sources for a firm's R&D. One source is its internal resources, commonly referred to as self-financed or private R&D; and the other source is governmental R&D, commonly referred to as public R&D. Simply disaggregating the *RD* variable in equation (4) to account for these two sources may not be the most appropriate way to think of the impact of public/private partnerships

---

[4] See, for example, Link and Siegel (2003) for the derivation of equation (4).
[5] There is a large empirical literature regarding estimates of ρ. It is also reviewed in Link and Siegel (2003).

on firm productivity.[6] Reflecting on the taxonomy of public/private partnerships in Table 1.1 in Chapter 1, one evaluation issue, which will be discussed in more detail in later chapters, is whether governmental involvement through a partnership relationship necessarily leverages private R&D, meaning does it increase the productivity of private R&D. Simply stated, is an estimate of ρ from equation (4) greater as a result of the partnership, or not?

A number of specific public/private partnerships will be discussed in later chapters. Tax incentives are discussed in Chapter 7, research collaborations in Chapter 8, and research joint ventures in Chapter 9. The enabling legislations for these public/private partnerships came, in part, in response to the significant TFP slowdown in the early 1980s, as seen in Figure 4.1.[7]

In retrospect, the Department of Commerce (1990) has documented that during the 1980s the United States began to lose its competitive advantage to Japan in a number of critical technology products: emerging materials, emerging electronics and information systems, emerging manufacturing systems, and emerging life-science applications. And, the United States began to lose it competitive advantage to the European Community in emerging manufacturing systems and emerging life-science applications. These competitiveness trends are also discussed in Chapter 10.

As the Department of Commerce (1990, p. 47) noted:

> As a [N]ation ... we no longer are totally self-sufficient in all essential materials or industries required to maintain a strong national defense. Consequently, we must identify requirements carefully and assess them against our industrial base capabilities. We must develop [R&D-based] strategies that enable us to meet security needs ... .

Also, as suggested in Chapter 2, there were two strategic thrusts for federal science and technology policy after World War II. One thrust was

---

[6] David, Hall, and Toole (2000) have reviewed the empirical literature in an effort to weigh the findings of the literature regarding whether public R&D is a complement to private R&D or a substitute. They conclude, absent overriding criticisms about the estimation methods and procedures used, that the findings are ambivalent.

[7] A detailed discussion of the myriad culprits associated with the TFP slowdown is in Link (1987) and Link and Siegel (2003).

the support of basic science, as evidenced by the National Science Foundation's creation and activities. The other was the support of public needs as defined by Congress, often called mission research (e.g., the development of advanced weapons systems by the Department of Defense). With the TFP slowdown in the early 1980s and subsequent decline in technology-based global competitiveness, the effectiveness of mission research was called into question and more emphasis was given to government's role to support industrial R&D.

DIMENSIONS OF R&D

Advancements in science and technology are drivers of technological change as reflected, in large part, in TFP growth. And, investments in R&D are a key indicator of advancements in science and technology. While this relationship between R&D and technological change is as important at the microeconomic level of firm behavior as it is at that macroeconomic level of economic growth. Thus, R&D is an important policy tool directly related to economic growth and secondarily related to global competitiveness.

For purposes of understanding the measurement of R&D, there are three fundamental dimensions. The first relates to the *source* of funding of R&D (who finances the investment), the second to the *performance* of R&D (who conducts the research and development), and the third to the *character of use* of R&D (whether the undertaking is of a basic or applied nature, or development). These three fundamental dimensions are not mutually exclusive.

Sources of Funding of R&D

The top row of Table 4.1 shows the sources for the nearly $313.4 billion of R&D expenditures in the United States in 2004. Industry accounted for nearly 64 percent of those expenditures; the federal government another 30 percent; and all other sources, universities and colleges, accounted for about 6 percent.

The primacy of industry in funding R&D has not always held, as shown in Figure 4.2. In the aftermath of World War II, up through the early 1980s, the federal government was the leading source of R&D funds in the Nation. Although the federal government was involved in supporting R&D before then, during the war the federal government dramatically expanded its R&D effort by establishing a network of federal laboratories, including atomic weapons laboratories. It was at that time that the federal government also greatly increased its support to extramural R&D performers, especially to a select group of universities and large industrial firms.

Table 4.1. National R&D Expenditures, by Selected Performer and Selected Funding Source, 2004 ($Millions)

Performer	Total R&D	Sources of R&D Funds				
		Federal government	Industry	Universities and colleges	Other non-profit institutions	Percent distribution, by performer*
TOTAL R&D	313,395	93,279	200,457	11,095	8,565	—
Federal government	24,807	24,807	—	—	—	7.9%
Industry	220,428	23,314	197,114	—	—	70.3%
Universities and colleges	42,431	26,115	2,135	11,095	3,087	13.5%
Other non-profit institutions	12,810	6,124	1,208	—	5,478	4.1%

* Percentages do not sum to 100.00 because not all performers are listed in the table (e.g., FFRDCs)
Source: National Science Board (2006).

After the war, federal R&D support continued to expand for both defense and non-defense purposes, including health R&D in the National Institutes of Health and—after the establishment of the National Science Foundation in 1950—a broad portfolio of basic research activities. As a result of a post-Sputnik national commitment to catch up to the Soviet space successes, federal support for space-related R&D mushroomed in the late 1950s and early 1960s. By 1960, the federal government accounted for 65 percent of the nation's total investment (80 percent of which was for defense), and industry accounted for nearly 33 percent of the total.

Figure 4.2. U.S. R&D Funding by Percentage Source of Funds, 1953-2004

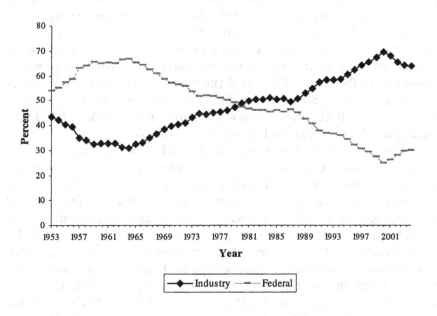

Source: National Science Board (2006).

Over the next twenty years the federal government continued to be the leading source of R&D funding, although the direction of its funding shifted over time. In the early 1960s, the relative defense share of federal R&D funding dropped precipitously from 80 percent in 1960 to about 50 percent in 1965, where it fluctuated narrowly until 1980. Early on, R&D for space exploration was the primary non-defense recipient of federal R&D funding. Indeed, more than three-fourths of federal non-defense R&D funds were in support of NASA's mission activities by 1965.

By 1970, however, after the success of several lunar landings, support for other non-defense purposes began to claim an increasingly larger share of the federal R&D totals, and continued to do so throughout the 1970s; notably growth in federal energy R&D occurred as a response to the several oil embargoes. Also by 1970, R&D support from industry was on the rise, and it accounted for just over 40 percent of the total national R&D

effort. As a result of relatively flat federal funding in the 1970s and continual slow growth from the industrial sector, the federal government and industry accounted for about equal shares by the early 1980s.

Since then, the federal government's share of R&D decreased to about 40 percent of the total in 1990 to its current 2004 share, just below 30 percent. Initially, the decreasing federal share came about even though federal dollar support for R&D—in absolute terms—was increasing.

Between 1980 and 1987, federal R&D rose about 40 percent after adjusting for inflation. Most of this growth, however, was in support of defense activities so that by 1987, the defense R&D share had grown to two-thirds of the federal R&D total (its highest share since 1963). After the break-up of the Soviet Union, the imperative for continual growth in federal defense R&D support was not as strong and the federal R&D total once again slowed (and even fell in constant dollars).

In terms of which agencies provide the R&D funds, federal sources are highly concentrated among just a few agencies. Five agencies accounted for 94.1 percent of all federal R&D funds: Department of Defense (47%), Department of Health and Human Services, primarily the National Institutes of Health (26%), National Aeronautics and Space Administration (9%), Department of Energy (8%), and National Science Foundation (4%).

Concurrent with recent reductions in federal R&D spending, major changes have also transpired in industrial R&D spending patterns. After lackluster funding in the early 1990s (reflecting the impact of mild economic recessions on its R&D activities) industry R&D support has grown rapidly since 1994 to almost 70 percent of the national R&D total in 2000, and is now (2004) 64 percent. As a result, and compared with the funding patterns of the mid-1960s, industry and government have reversed positions.

R&D Performers

R&D is performed in what has been called the U.S. national innovation system (Nelson 1993). The venue for the performance of R&D within the system—and this is true in all industrial nations—are research laboratories. Many scholars have set forth alternative definitions of a national innovation system; Crow and Bozeman (1998, p. 42), as one example, write that the U.S. national innovation system may be thought of as:

> ... the complex network of agents, policies, and
> institutions supporting the process of technical advance in
> an economy.

The laboratory performers of R&D correspond to the sectors that finance R&D, but not all R&D funded by a sector is performed in that sector. For example, industry performed in its laboratories approximately $220 billion of R&D in 2004, of which $197 billion came from industry itself. The additional amount of R&D performed by industry came from the federal government. See Table 4.1.

Almost one-fourth of the R&D funded by the federal government is performed in industry, and more than one-half of these funds are spent in the aircraft, missiles, and transportation equipment industries. Universities and colleges fund only about 26 percent of the R&D they perform in their laboratories. Nearly 62 percent of R&D performed in universities and colleges comes from the federal government and the rest equally from industry, nonprofit institutions, and nonfederal government sources.

Figure 4.2 shows that since the late-1980s the rate of decline in federal funding of national R&D has increased. The lion's share of that decrease has come in the form of federal allocations for R&D performed in industry, for which the R&D level of support displays a roller-coaster-like pattern. The latest peak in federal support for industrial R&D was due to major defense-related funding increases for President Reagan's Strategic Defense Initiative, prior to the collapse of the Soviet Union. By contrast, federal funding to universities and colleges, adjusted for inflation, has increased slightly each year since at least the late 1970s.

There are other important dimensions to the performance of industrial R&D. Approximately three-fourths of total industrial R&D is performed in manufacturing industries. The dominant manufacturing industries in terms of dollars of R&D performed are chemicals and allied products, electrical equipment (including computers), and transportation equipment. The remaining one-fourth is performed in the non-manufacturing sector, including services. Computer-related services are the leaders therein. The steep growth of R&D performed in the services is a relatively recent phenomenon. As recently as 15 years ago, manufacturers still accounted for more than 90 percent of total industrial R&D. Now (2004), manufacturing accounts for about 60 percent of total industrial R&D.

Also, not all industry-performed R&D occurs within the geographical boundaries of the United States. Of the nearly $195 billion in R&D performed by industry in 2002 (the latest year for which the foreign-performed data are available), about $21 million, or almost 11 percent, were conducted in other countries.

Foreign investments in R&D are not unique to U.S. firms; the outflow of U.S. industrial R&D into other countries is approximately offset by an inflow of others' R&D to be performed in the United States. Most (almost 70%) of U.S.-funded R&D abroad was performed in Europe—primarily in Germany, the United Kingdom, and France.

Overall, U.S. R&D investments abroad have gradually shifted away from the larger European countries, and Canada, and toward Japan, several of the smaller European countries (notably Sweden and the Netherlands), Australia, and Brazil. Pharmaceutical companies accounted for the largest industry share, much of which took place in the United Kingdom.

Substantial R&D investments are made by foreign firms in the United States. In 2002, R&D investments were concentrated in drugs and medicines (mostly from Swiss, German, and British firms), industrial chemicals (funded predominantly by German and Dutch firms), and electrical equipment (one-third of which came from French affiliates).

R&D BY CHARACTER OF USE

Vannevar Bush is credited for first using the term "basic research." In his 1945 report to President Roosevelt, *Science—the Endless Frontier*, Bush used the term and defined it to mean research conducted without thought of practical ends. Since then, policy makers have been concerned about definitions that appropriately characterize the various aspects of scientific inquiry that broadly fall under the label of R&D and that relate to the linear model that Bush proffered.

Definitions are important to the National Science Foundation because it collects expenditure data on R&D. For those data to accurately reflect industrial and academic investments in technological advancement, and for those data to be comparable over time, there must be a consistent set of reporting definitions.

The classification scheme used by the National Science Foundation for reporting purposes was developed for its first industrial survey in 1953-1954, as documented in Link's (1996b) history of the classification scheme. While minor definitional changes were made in the early years, namely to modify the category originally referred to as "basic or fundamental research" to simply "basic research," the concepts of basic research, applied research, and development have remained much as was implicitly contained in Bush's 1945 linear model:

Basic Research → Applied Research → Development

The objective of *basic research* is to gain more comprehensive knowledge or understanding of the subject under study, without specific applications in mind. Basic research is defined as research that advances scientific knowledge but does not have specific immediate commercial objectives, although it may be in fields of present or potential commercial interest. Much of the scientific research that takes place at universities is basic research.

Applied research is aimed at gaining the knowledge or understanding to meet a specific recognized need. Applied research includes investigations oriented to discovering new scientific knowledge that has specific commercial objectives with respect to products, processes, or services.

Development is the systematic use of the knowledge or understanding gained from research directed toward the production of useful materials, devices, systems, or methods, including the design and development of prototypes and processes.

Approximately 62 percent of national R&D is development, with almost 19 percent of R&D being allocated to applied research and the same approximate percentage to basic research. Different sectors contribute disproportionately to the Nation's funding and performance of these R&D component categories. Applied research and development activities are primarily funded by industry and performed by industry. Basic research, however, is primarily funded by the federal government and generally performed in universities and colleges.

Table 4.2 shows the distribution of R&D by character of use and by source of funds for 2004. Industry funds the 84 percent of its own basic and over 88 percent of its applied research; the federal government funds nearly 65 percent of basic research at universities and colleges and 54 percent of the applied research performed there. Nearly 89 percent of all development is funded by industry and performed by industry.

The decline in federal support of R&D over the past two decades—see Figure 4.2—has primarily come at the expense of applied research and development performed in industry.

Table 4.2. National R&D Expenditures by Character of Use, Performer,
and Source, 2004: 2004 ($Millions)

		Source of Funds		
Performer	Total R&D	Industry	Federal government	Universities and colleges
Basic Research	57,711			
Industry	8,514	84.0%	13.4%	—
Federal government	4,973	—	100%	—
Universities and colleges	31,735	4.6%	64.9%	23.9%
Applied Research	57,150			
Industry	31,795	88.2%	11.8%	—
Federal government	8,415	—	100.0%	—
Universities and colleges	9,223	0.6%	54.0%	31.3%
Development	194,065			
Industry	175,690	88.6%	11.4%	—
Federal government	11,419	—	100.0%	—
Universities and colleges	1,474	8.3%	36.8%	42.9%

Source: National Science Board (2006).

5 ALTERNATIVE MODELS OF TECHNOLOGICAL CHANGE

The production function model of technological change set forth in Chapter 4 has guided for over one-half of a century much of the empirical research in economics related to the relationship between R&D and technological change. Simply put, its underlying conceptual framework is:

R&D → Knowledge → Innovation → Technological Advancement → Economic Growth

Thus, conceptually, there is a positive relationship between investments in R&D and technological change as measured by growth over time in TFP, and statistically the correlation is also positive.[1]

However, the production function approach that has dominated the empirical economics of this topic is void of any statement about the role of innovation in the R&D-to-TFP relationship, thus a rethinking of technological change, as well as the idea that many factors may be causally related is warranted.

TECHNOLOGY AND TECHNOLOGICAL CHANGE[2]

Researchers have used the concept of technology in a variety of ways. In a narrow sense, technology refers to specific physical or tangible tools, but in a broader sense technology describes whole social processes. In the broader sense, technology refers to intangible tools. Although there are analytical advantages to both the narrow and the more encompassing

---

[1] This literature is reviewed in Link (1987) and Link and Siegel (2003).

[2] This section draws, in part, from Bozeman and Link (1983), Link (1987), Hébert and Link (1988), and Link and Siegel (2003).

views, the different uses of the concept of technology invariably promote confusion at both the theoretical, empirical, and policy levels.

By focusing on physical or tangible technology, questions arise such as: How can technologies be differentiated? What aspects of technology are of interest? For the most part, economists have attempted to answer such questions by dealing with the indirectly perceivable aspects of physical technology or tangible tools. Namely, the focus turns from attributes to the knowledge embodied within the technology. And, the knowledge base of technology is not only a theme in this chapter, but also it is a critically important starting point for the development of science and technology policy.

Conceptualizing technology as the physical representation of knowledge provides a useful foundation for understanding technological change and its determinants. Any useful device is, in part, proof of the knowledge-based or informational assumptions that resulted in its creation. The information embodied in a technology varies accordingly to its source, its type, and its application. For example, one source of information is science, although scientific knowledge is rarely sufficient for the more particular needs entailed in constructing, literally, a technological device. Having said that, it would be useful in this regard to think of science as focusing on the understanding of knowledge and technology as focusing on the application of knowledge.

Other sources of knowledge include information from controlled and random experimentation, information that philosophers refer to as ordinary knowledge, and finally, information of the kind that falls under the rubrics of creativity, perceptiveness, and inspiration.

Regarding perceptiveness, Machlup (1980, p. 179) argued that formal education is only one form of knowledge. He asserted that knowledge is also gained experientially and is gathered and processed at different rates by each individual. The following statement reflects Machlup's notion of perception quite clearly:

> Some alert and quick-minded persons, by keeping their eyes and ears open for new facts and theories, discoveries and opportunities, perceive what normal people of lesser alertness and perceptiveness, would fail to notice. Hence new knowledge is available at little or no cost to those who are on the lookout, full of curiosity, and bright enough not to miss their chances.

Machlup's informational view of technology implies that technology *per se* is an output that arises from a formal, rational, purposively undertaken process. Such an idea—the production of technology—highlights the role of research in the generation of technologies. And, the concept of research underscores the myriad sources available from which knowledge can be acquired, formal R&D being one source. Technologies can thus be distinguished, albeit imperfectly, by the amount of embedded information. More concretely, R&D activities—wherever they are based—play a large role in creating and characterizing new technologies.

Closely related to the concept of technology is the notion of invention and innovation. Following Bozeman and Link (1983, p. 4):

> The concepts commonly used in connection with innovation are deceptively simple. *Invention* is the creation of something new. An invention becomes an *innovation* when it is put in use.

It is useful to think of an innovation as something new that has been brought into use.[3] Thus, the innovation represents, in a sense, a new underlying technology. When the innovation is itself the final marketable result, it is sometimes referred to as a product innovation. When the innovation is applied in a subsequent production process, it is sometimes referred to as a process innovation (meaning that its application affects a production process).

Embedded in this distinction between invention and innovation is a process whereby inventions become applied. This process is often referred to as the innovation process, with an innovation being the defined result or output of the process.

For the purposes of a broader view of technological change, it could be useful to incorporate explicitly the above concepts of invention and innovation as well as the concept of entrepreneurship. The contributions of Schumpeter are a starting point. According to Schumpeter (1939, p. 62) innovation can meaningfully be defined in terms of a production function, and, in a sense, as a factor shifting the production function (as modeled in Chapter 4):

---

[3] The characteristics of newness appear in the writing of many scholars who attempted to address this topic. Kuznets (1962, p. 19), for example, refers to inventive activity as a "new combination of available knowledge."

> [The production function] describes the way in which
> quantity of product varies if quantities of factors vary. If,
> instead of quantities of factors, we vary the form of the
> function, we have an innovation.

Schumpeter noted that mere cost reducing applications of knowledge lead
only to new supply schedules of existing goods. Therefore, this kind of
innovation must involve a new commodity or one of a higher quality. This
is what economists typically refer to as product innovation. He also noted
that the knowledge undergirding the innovation need not be new; it may be
existing knowledge that has not been utilized before. According to
Schumpeter (1928, p. 378):

> [T]here never has been anytime when the store of
> scientific knowledge has yielded all it could in the way of
> industrial improvement, and, on the other hand, it is not
> the knowledge that matters, but the successful solution of
> the task *sui generis* of putting an untried method into
> practice—there may be, and often is, no scientific novelty
> involved at all, and even if it be involved, this does not
> make any difference to the nature of the process.

Driving this process is the entrepreneur.[4]

THE ENTREPRENEUR AND ENTREPRENEURSHIP

The concept of the entrepreneur can be traced at least as far back as the
Physiocrats in France in the mid-1700s. Nicolas Baudeau (1910, p. 46)
referred to a process guided by an active agent, which he called an
entrepreneur, within a capitalistic system:

---

[4] More than two decades after these writings of Schumpeter, Usher (1954)
independently rediscovered these same concepts. He posited that technology is the
result of an innovation, and an innovation is the result of an invention. An
invention, of course, results as the emergence of new things requiring an act of
entrepreneurial insight going beyond the normal exercise of technical or
professional skills.

> Such is the goal of the grand productive enterprises: first
> to increase the harvest by two, three, four, ten times if
> possible; secondly to reduce the amount of labor
> employed and so reduce costs by a half, a third, a fourth,
> or a tenth, whatever possible.

Embedded in this conceptualization of entrepreneurship is the notion of an innovative process, one perhaps as simple as the perception of new technology adopted from others so as to increase agricultural yield, or one as refined as the actual development of a new technology to do the same. When the process is completed and when the innovation is put into use there will be a productivity gain and possibly even a substitution of capital for labor.

Entrepreneurship is a process: an output of that process is the promotion of one's own innovation or the adoption of another's innovation. For the purposes of posting an alternative model of technological change, the term entrepreneur is defined, following Hébert and Link (1988), as one who perceives an opportunity and has the ability to act upon it. Hence, entrepreneurship, much like innovation which is guided by entrepreneurial action, is a process that involves both *perception* and *action*. (More historical background on the entrepreneur—who he is— and entrepreneurship—what he does—is in an appendix to this chapter.)

From the perspective of a technology-based firm or organization, the perception of an opportunity may be influenced by changes in strategic directions or competitive markets, but perception of an opportunity is fundamentally only the first step. The consequent step is the ability to act on that perception. What defines the entrepreneur is the ability to move invention forward into innovation. The invention may be discovered or developed by others. The entrepreneur is able to recognize the commercial potential of the invention and organize the capital, talent, and other resources that turn an invention into an innovation and then into a commercially viable technology.

There are several requisite resources needed for action, and action takes the perception of an opportunity forward to result in an innovation. One obvious and fundamental resource is R&D because it not only provides a stock of knowledge to encourage perception but also the ability for the firm to foster action. However, firms that do not conduct R&D can still be entrepreneurial. In such firms, innovations are likely to be introduced rather than produced. Such firms act in an entrepreneurial manner by hiring creative individuals and providing them with an environment conducive for the blossoming of their talents.

Consider R&D-active firms. The R&D they conduct serves two general purposes. First, it provides the resource base from which the firm can respond to an opportunity with perceived strategic merit or technical opportunity. This opportunity allows the firm to develop a commercial market. Second, those scientists involved in R&D are the internal resource that facilitates the firm's being able to make decisions regarding the technical merits of others' innovations and how effectively those innovations will interface with the existing technological environment of the firm.

The firm may choose to purchase or license this technology or undertake a new R&D endeavor. In this latter sense, one related and very important role of R&D is to enhance the absorptive capacity of the firm (Cohen and Levinthal 1989).

Thus, the role of R&D as enhancing the absorptive capacity of the firm goes beyond simply assessing the technical merits of potentially purchasable technology. It allows the firm to interpret the extant technical literature; to interface when necessary with the research laboratories of others, in a research partnership relationship (discussed below) or to acquire technical explanations from, say, a federal laboratory or university laboratory; or simply to solve internal technical problems.[5]

ALTERNATIVE MODELS OF TECHNOLOGICAL CHANGE

Two models of innovation are set forth in this section. The first relates to a technology-based manufacturing sector firm and the second to a technology-based service sector firm.[6] Based on the discussion in Chapter 4 about trends in service sector R&D, both sectors are thus important to technological change.

Each model contains a relationship between entrepreneurship, innovation, R&D, and technological change, key elements that were missing from the traditional production function model in Chapter 4.

---

[5] It is not surprising that public policies aimed at enhancing the innovativeness of firms, especially in light of increasing global competition as was the case beginning in the late 1970s and early 1980s, have focused directly or indirectly on R&D. These public policies are discussed in later chapters.

[6] These models come from Gallaher et al. (forthcoming).

Model for a Manufacturing Sector

At the root of the model in Figure 5.1 is the science base, referring to the accumulation of scientific and technological knowledge. The science base resides in the public domain. Investment in the science base comes from basic research, primarily funded by the government and primarily performed globally in universities and federal laboratories.

Figure 5.1. Entrepreneurial Model of Innovation in a Technology-based Manufacturing Sector Firm

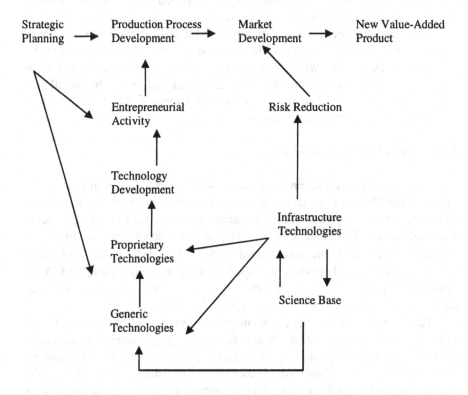

For a technology-based manufacturing firm, technology development in the form of basic and applied research generally begins within its laboratory. There, R&D involved the application of scientific knowledge toward the proof of concept of a new technology. Such fundamental

research, if successful, yields a prototype or generic technology. If the prototype technology has potential commercial value, follow-on applied research takes place toward development. If successful, a proprietary technology will result.

Basic research, applied research, and development—ala Vannevar Bush's linear model—occur within the firm as a result of its strategic planning and guide its entrepreneurial activities. Generally, strategic planning involves the formulation of road maps for developing new emerging technologies. A manufacturing firm targets discrete technology jumps, creating new technologies that make their competition obsolete, their strategic plans are long term and not closely linked to current competitive planning. Entrepreneurial activity then drives the firm toward the production of the new product or process.

Infrastructure technologies emanate from the science base. These technologies, such as test methods or measurement standards, reduce the market risk associated with the market introduction of a new product or process. Once a new product has been designed and tested, technical risk may be low, but market risk may be significant until the product is accepted and integrated into existing systems.

Model for a Service Sector Firm

The model for the technology-based service sector firm in Figure 5.2 differs from that of the manufacturing sector firm in several dimensions. While innovation in both the manufacturing firm and the service sector firm builds on the science base, the service sector firm is likely to acquire—purchase (Link and Zmud 1987) or license (Link and Scott 2002)—products and services as inputs that incorporate others' R&D, as opposed to conducting it internally. And, of course, those from whom such acquisitions are made draw on the science base.

The service sector firm's strategic planning focuses primarily on retaining or gaining market shares for existing products and services. Thus, innovation is guided by customer input and competitive planning that involves continual or incremental transition strategies. The entrepreneurial activity of the service sector firm drives redesigned or reconfigured enhancements of its existing products. At the root are other's technologies that are licensed or purchased to meet the firm's road map for deploying modifications of its existing products. This product and service enhancement often involve systems integration where systems integration facilitates the intersection of hardware, software, and the synthesis of application domains such as finance, manufacturing, transportation, and retail.

For the service sector firm, infrastructure technologies ensure that purchased technologies interface or integrate with the service sector firm's existing systems. Such infrastructure technologies also emanate from the science base.

Risk reduction is also an important element in the model of innovation in a service sector firm. It is more likely to be less than in a manufacturing sector firm because the service sector firm's innovation is enhancing products in existing markets, assuming that technical risk is minimal.

Figure 5.2. Entrepreneurial Model of Innovation in a Technology-based Service Sector Firm

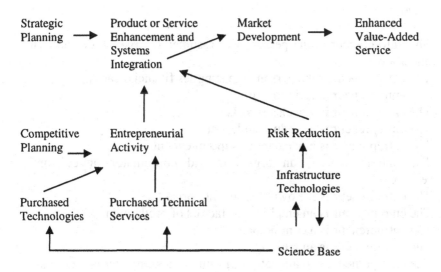

APPENDIX ON
ENTREPRENEURSHIP

The history of economic thought holds diverse opinions about the nature and role of the entrepreneur.[7]   At least twelve distinct themes have appeared within the economics literature about the entrepreneur. These themes are:

1. The entrepreneur is the person who assumes the risk associated with uncertainty.
2. The entrepreneur is the person who supplies financial capital.
3. The entrepreneur is an innovator.
4. The entrepreneur is a decision maker.
5. The entrepreneur is an industrial leader.
6. The entrepreneur is a manager or superintendent.
7. The entrepreneur is an organizer and coordinator of economic resources.
8. The entrepreneur is the owner of an enterprise.
9. The entrepreneur is an employer of factors of production.
10. The entrepreneur is a contractor.
11. The entrepreneur is an arbitrageur.
12. The entrepreneur is an allocator of resources among alternative uses.

With regard to the twelve themes above, statements 2, 6, 8, and 9 describe static roles for the entrepreneur. In a static world, the entrepreneur is a passive element because his actions merely constitute repetitions of past procedures and techniques already learned and implemented.   The   prevailing   wisdom   by   those   who   study

---

[7] This section draws from Hébert and Link (1988, 1989).

entrepreneurship from an evolutionary perspective eschew his static role. Only in a dynamic world does the entrepreneur become a robust figure, and a dynamic role for him is implied in the other eight thematic statements. But even then, the innovative activities of small firms are not explicit.

To place the entrepreneur, and his dynamic entrepreneurial actions, in an economics concept, one must begin with the ideas proffered by Cantillon (c. 1680 – 1734). From Cantillon, three intellectual branches loosely referred to as the German Tradition characterized by, among others, von Thünen (1785 – 1850) and Schumpeter (1883 – 1950); the Chicago School characterized by, among others, Knight (1885 – 1972) and Schultz (1902 – 1998); and the Austrian Tradition characterized by, among others, von Mises (1881 – 1973) and Shackle (1903 – 1992). The representative contributions of Cantillon, Schumpeter, Schultz and von Mises are briefly discussed below.

Richard Cantillon

In the 18th century, Cantillon outlined the framework of a nascent market economy founded on individual property rights and based on economic interdependency. He recognized three classes of agents: landlords, who are financially independent; entrepreneurs, who engage in market exchanges at their own risk in order to make a profit; and hirelings, who work for fixed wages.

Cantillon's entrepreneur is, within his framework, the central economic actor. He is someone who engages in exchanges, in the face of uncertainty, for profit. What Cantillon stressed is the function of the entrepreneur rather than his social status. Those who take chances in hopes of gain are entrepreneurs; even beggars and robbers who face uncertainty are entrepreneurial. It is the lack of perfect foresight that is the origin of the entrepreneurial spirit. If Cantillon's entrepreneur constantly had to exercise business judgment, and if he guessed wrong he must be accountable for the result. This Cantillonesque notion of the entrepreneur would later be widened by both Knight and von Mises.

Joseph Schumpeter

The concept of the entrepreneur was unveiled by Schumpeter within the context of economic development. Quite simply, the entrepreneur was the *persona causa* of economic development; he was the mechanism of economic change.

To Schumpeter, competition involved mainly the dynamic innovations of the entrepreneur. Schumpeter used the theoretical concept of equilibrium, "the circular flow of economic life," as his point of departure. Economic life proceeds routinely on the basis of past experience, he argued. In this equilibrium state, the entrepreneur is a non-entity because the relationship between inputs and production is invariant. The real question about economic growth and development is not how capitalism administers existing structures but how capitalism creates and destroys them. This process—"creative destruction"—is the essence of economic development. In other words, development is the disturbance of the circular flow, and the process of carrying out new combinations of inputs in production is done by, or is the role of, the entrepreneur.

T.W. Schultz

Schultz's theory of entrepreneurship is rooted in the economic theory of human capital. His points of departure rest on four criticisms of the earlier treatment of the entrepreneur: (1) the concept is generally restricted to businessmen, (2) it does not take into account the differences in allocative abilities among entrepreneurs, (3) the supply of entrepreneurship is not treated as a scarce resource, and (4) entrepreneurship is neglected in favor of general equilibrium considerations.

Schultz refined the concept of entrepreneurship to be the ability to deal with disequilibria, and he extended the notion of entrepreneurial activity to include non-market activities (e.g., household decisions, allocation of time) as well as market activities. He also produced evidence of the effects of education on individuals' ability to perceive and react to disequilibria. Schultz argued that Schumpeter's conceptualization of entrepreneurship did not go far enough because it was confined to development disequilibria.

Ludwig von Mises

According to von Mises, economics is the study of human action, and human action that is distinctively economic takes place within a market framework. The nature of market activity is that it is an entrepreneurial process. The fundamental aspect of von Misesian human action is that it influences the future and is influenced by the future. Thus, participants in market activity make choices and cope with the subsequent uncertainties of the future. Within this von Misesian framework, every man is an entrepreneur because every man makes choices (or decisions) and is then subject to the uncertainty that they create.

6 THE PATENT SYSTEM

The U.S. patent system is a public/private partnership.[1] As discussed in Chapter 1, a public/private partnership is an innovation-related relationship that involves public and/or private resources. The patent system involves both public and private resources in its maintenance.

In terms of the taxonomy used to characterize public/private partnerships, the patent system, promulgated through the Patent Act of 1790, is an example of indirect governmental involvement, because the system is in place and it provides an innovative environment in which firms can optimize, and the economic objective of the patent system is to leverage private R&D. See Table 6.1. The patent system is in the cell corresponding to indirect governmental involvement with an economic objective to leverage private R&D.

In subsequent chapters, various public/private partnerships will be placed in the taxonomy described by Table 6.1.

HISTORY OF THE U.S. PATENT SYSTEM

The history of the U.S. patent system dates to the authority given to Congress in the Constitution of the United States. Article I, section 8 states:

> Congress shall have power ... to promote the progress of science and useful arts, by securing for limited times to

---

[1] This chapter is based on Link (1999b). Link (1999b) was later expanded in Audretsch et al. (2002a) and then reproduced in book form as Feldman, Link, and Siegel (2002).

authors and inventors the exclusive right to their
respective writings and discoveries.

Based on this authority, Congress initiated a number of patent laws
beginning in 1790 with the Patent Act.[2] The version of law that is now in
effect was enacted on July 19, 1952 (to be effective January 1, 1953).

Table 6.1. Taxonomy of Public/Private Partnerships

	Economic Objective	
Governmental Involvement	*Leverage Public R&D*	*Leverage Private R&D*
Indirect		**Patent system**
		(Patent Act)
Direct		
Financial Resources		
Infrastructural Resources		
Research Resources		

The Patent and Trademark Office issues patents for inventions. The
patent term is 20 years, and it grants exclusive property rights to the
inventor over that period of time. Patents are effective only within the
United States and its territories and possessions.

U.S. patent law is clear about what can be patented. Any person who:

> ... invents or discovers any new or useful process,
> machine, manufacture, or composition of matter, or any
> new and useful improvement thereof, may obtain a patent.

It is important to note the word "useful," recalling that Franklin created the
American Philosophical Society of Philadelphia in 1742 to promote
"useful knowledge."

---

[2] The Patent Act of 1790 was influenced by President Thomas Jefferson, among
others. The concepts therein trace to English law where, as precedence, in 1449,
King Henry VI awarded a patent to John of Utynam for stained glass
manufacturing. A readable overview of the history of the U.S. Patent Office is in
Watson (2001).

Three criteria must hold to be granted a patent: utility, novelty, and non-obviousness. Utility means that the invention must be useful; novelty means that the invention must be new and not merely a copy or repetition of another invention; and, non-obviousness—the most difficult criterion—means that the invention must neither be suggested by previous work nor totally anticipated given existing practices.

While the U.S. Code applies to patents granted in the United States, and in the territories and possessions of the United States, treaties have been promulgated to extend protection beyond national boundaries. The Paris Convention for the Protection of Intellectual Property of 1883 provided that each of the 140 signatory nations recognized the patent rights of other countries. Subsequent treaties have extended such coverage and made filings in other countries more efficient.

THE ECONOMICS OF PATENTING

Figure 6.1 illustrates what may be called the economics of patenting from the perspective of the firm.[3] The marginal private rate of return to R&D is measured on the vertical axis and the level of R&D spending is measured on the horizontal axis. The marginal private return schedules are downward sloping reflecting diminishing returns to R&D in any given time period, and for simplicity we assume the marginal private return schedule to be linear. Absent the patent system, the firm will choose to invest RD_0 in R&D. This is an optimal investment for the firm; it invests to the point where its marginal private cost of R&D (assumed to be constant for simplicity) equals its marginal private return. However, assume that the project society wants the firm to invest in requires $RD_1 > RD_0$.

Level RD_1 in Figure 6.1 is, for illustration purposes, sufficient for the socially desirable project B, say, in Figure 3.1 to be undertaken, but the firm does not have an incentive to invest in R&D at that level. The key point is that the existence of the patent, or more precisely, the expectation that the firm will be awarded a patent, can induce the firm to devote additional resources to R&D (to reach level RD_1.)

Receipt of monopoly power for 20 years through a patent increases the firm's marginal private return from its investments in R&D, thus shifting the marginal private return schedule to the right. The intersection of the

---

[3] This model benefited from discussions with Bronwyn Hall.

new marginal private schedule and the marginal private cost schedule defines the optimal investment level for the firm; it is RD_1.

Figure 6.1. Economics of Patenting: Increasing Marginal Private Return for the Firm

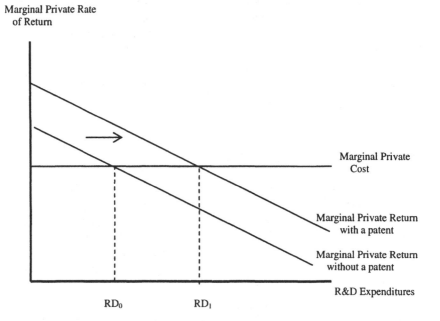

Researchers have investigated a number of economic issues related to patenting activity, and the economic role of patents in the innovation process. Some significant findings from this body of research are:

- There is a strong positive correlation at the firm level between R&D expenditure (or employment) and patents.
- There is a positive correlation between patenting activity and the market value of the firm.
- The economic value of patents is highly skewed, where value is determined by citations.
- A citation-weighted measure of patents is more highly correlated with market value of the firm than is an unweighted measure.

- Patents are the primary instrument used by firms to protect their intellectual capital.

The literature on patents is clear that a simple count of patents is not the best predictor of a nation's or a firm's innovative activity, much less its technological advancement.

TRENDS IN PATENTING

Figure 6.2 shows utility patent applications and patents granted in the United States from 1953 through 2003. Several trends in patenting activity in the United States are noteworthy:

- In the early 1980s, the number of patents awarded to U.S. inventors began to decline and the number of U.S. patents awarded to foreign inventors began to rise, thus causing some policy makers to question the inventive environment in U.S. firms. This trend was yet another indicator that U.S. global competitiveness was declining at that time.
- During the 1980s, the largest number of U.S. patents awarded to foreign inventors was granted to Japanese inventors. In fact, in 1995, over 20,000 patents were awarded to Japanese inventors, compared to about 7,000 for the next highest represented country, Germany.
- The share of total patents awarded to foreign inventors is low in the United States compared to other countries. It is highest in Italy and Canada and lowest in Japan and Russia.
- During the past decade, Japanese inventors have more international patents in three important technologies than inventors from any country, with the United States being second. These technologies are: robotics, genetic engineering, and advanced ceramics.
- Since the enactment of the Bayh-Dole Act of 1980 (see Chapter 8), which transferred ownership of intellectual property from federal agencies to universities, there has been a rapid rise in university patenting (Henderson, Jaffe, and Trajtenberg 1998).

According to Kortum and Lerner (1999), conventional wisdom held that the Court of Appeals of the Federal Circuit, created in 1982 by the Federal Court Improvements Act—created to reconcile patent disputes efficiently—was the driving force for the increase in patent activity shown in Figure 6.2.[4] But, their empirical research suggests that the increase was due instead to improvements in industrial firms' management of their R&D, and relatedly a shift toward more applied research.

Figure 6.2. Trends in Patenting in the United States

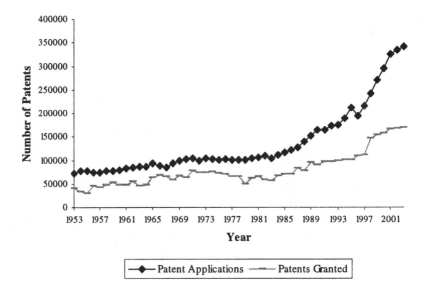

---

[4] According to Hall (2005), after 1983-1984, patent activity increased significantly, but primarily in the electrical and computing technology sectors.

7 TAX INCENTIVES

$\mathcal{T}$ax incentives represent a public/private partnership.[1] In terms of the taxonomy used to characterize public/private partnerships, tax incentives, and the legislation that promulgates them, represent indirect governmental involvement in innovation. The economic objective of tax incentives, in particular the R&E tax credit, is to leverage private R&D. See Table 7.1.

TAX INCENTIVES

Tax incentives, in general, are a mechanism that government uses to stimulate or leverage private sector R&D. Like any policy tool, tax incentives have advantages and disadvantages. Advantages include the following (Bozeman and Link 1984):

- Tax incentives entail less interference in the marketplace than do other mechanisms, thus affording private-sector recipients the ability to retain autonomy regarding the use of the incentives.
- Tax incentives require less paperwork than other programs.
- Tax incentives obviate the need to directly target individual firms in need of assistance.
- Tax incentives have the psychological advantage of achieving a favorable industry reaction.
- Tax incentives may be permanent and thus do not require annual budget review.
- Tax incentives have a high degree of political feasibility.

---

[1] This chapter is based on Bozeman and Link (1984), various presentations by Link at the OECD in the early 1990s, and Link (1999b).

Some disadvantages of tax incentives are:

- Tax incentives may bring about unintended windfalls by rewarding firms for what they would have done in the absence of the incentive.
- Tax incentives often result in undesirable inequities.
- Tax incentives raid the federal treasury.
- Tax incentives frequently undermine public accountability.
- The effectiveness of tax incentives often varies over the business cycle.

Table 7.1. Taxonomy of Public/Private Partnerships

	Economic Objective	
Governmental Involvement	*Leverage Public R&D*	*Leverage Private R&D*
Indirect		Patent system (Patent Act)
		Tax incentives (R&E tax credit)

Direct
  Financial Resources
  Infrastructural Resources
  Research Resources		

THE ECONOMICS OF TAX CREDITS

Figure 7.1 illustrates the economics of a tax credit. The marginal rate of return is measured on the vertical axis and the level of R&D spending is measured on the horizontal axis. Both the marginal social return and the marginal private return schedules are downward sloping reflecting diminishing returns to R&D investments in a given time period. The social return schedule is drawn greater than the private return schedule for all levels of R&D because firms cannot appropriate all the benefits from conducting R&D; some of those benefits spillover to other firms in the current time period and in the post-innovation time period thus generating additional benefits to society. The marginal cost to the firm to undertake R&D is shown to be constant (horizontal).

Figure 7.1. Economics of a Tax Credit: Decreasing Marginal Private Cost for
the Firm

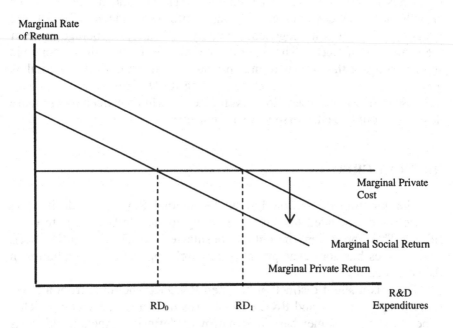

As drawn, the firm will equate the marginal private cost of conducting
R&D with the marginal private return associated with this activity, and the
firm will invest at the level RD_0. As in the patent example in Figure 6.1,
assume that society wishes to push the firm to an investment level RD_1 to
undertake a project like, say, B in Figure 3.1. Society, given the firm's
marginal cost schedule, would like the firm to invest in R&D to maximize
social benefits. Hence, the optimal tax credit is one that provides an
incentive to the firm to increase its R&D to point RD_1. Receipt of a tax
credit can be thought of as a reduction in the marginal private cost of
undertaking additional R&D, and the firm will re-equate its new marginal
private cost with its marginal private return and invest at RD_1.

Unlike patents, a tax credit on R&D simply increases the firm's
private return on marginal R&D projects by reducing its marginal private
cost to undertake such projects. Thus, tax incentives will increase the
firm's level of R&D from RD_0 to RD_1 but will not alleviate the technical
or market risk that characterizes the firm's portfolio of projects.

However, because R&D is not a homogeneous activity and because, as Mansfield (1980) and Link (1981) have shown, the research (R) portion of R&D has a greater impact on productivity growth and hence economic growth than does development (D), any uniform tax incentive that treats R&D as if it were a homogeneous activity will likely encourage more of the same mix of R&D. That may not necessarily be bad since economic studies suggest that the marginal private return from R&D in total is greater than the marginal cost of conducting R&D. However, a tax credit on research as opposed to development, while conceptually more desirable, could be cumbersome to administer.

R&E TAX CREDIT

The adoption of Section 174 of the Internal Revenue Code in 1954 codified and expanded tax laws pertaining to the R&D expenditures of firms. This provision permitted businesses to deduct fully R&E expenditures but not development or research application expenditures in the year incurred.

There is a slight distinction between R&D expenditures from a NSF-reporting perspective and R&E expenditures from a tax perspective. R&E expenditures are somewhat more narrowly defined to include all costs incident to development. R&E does not include ordinary testing or inspection of materials or products for quality control of those for efficiency studies, etc. R&E, in a sense, is the experimental portion of R&D. That said, in practice it is often difficult to distinguish one category from the other.

Under Section 174, businesses are not allowed to expense R&E related equipment. Such equipment must be depreciated. However, the Economic Recovery and Tax Act of 1981 (ERTA) provided for a faster depreciation of R&E capital assets than other business capital assets.

ERTA also included a 25 percent tax credit for qualified R&E expenditures in excess of the average amount spent during the previous three taxable years or 50 percent of the current year's expenditures (the R&E base). The initial R&E tax credit had several limitations including the fact that it did not cover expenses related to the administration of R&D or to research conducted outside of the United States. The Tax Reform Act of 1986 modified these limitations, but reduced the marginal rate from 25 percent to 20 percent. Over the years the credit had been modified, primarily in terms of the definition of the R&E base, but the credit has never been made permanent. It has expired a number of times, only to be renewed retroactively. President George W. Bush has advocated for the credit to be made permanent.

In 1996, the Congressional Office of Technology Assessment released a report on the effectiveness of the R&E tax credit. The report concluded:

- There is not sufficient information available to conduct a complete benefit-to-cost analysis of the effectiveness of the R&E tax credit on the economy.
- The econometric studies that have been done to date conclude that the credit has been effective in the sense that for every dollar lost in federal revenue there is an increase of a dollar in private sector R&D spending. These studies conclude that the credit would be more effective if it were made permanent. Hall and van Reenen (2000) conclude from their review of the literature that the tax elasticity of R&D is about unity, meaning that a one percent increase in the credit will increase industry R&D by about one percent.
- The R&E tax credit represents a small fraction of federal R&D expenditures, about 2.6 percent of total federal R&D funding and about 6.4 percent of federal R&D for industry.

The R&E tax credit is not unique to the United States (Leyden and Link 1993). Japan's tax credit is marginal, and it was initiated in 1966. Canada also initiated a program in the 1960s, but their program is a flat tax program.

Some (e.g., Link and Bauer, 1989) have proposed a tax credit for cooperative research. Economic theory concludes that firms that cooperate with each other in a research joint venture type of arrangement have the incentive to cooperate at the research end of the R&D spectrum rather than at the development end. Thus, a tax credit for cooperative research involvement is theoretically a viable alternative to the R&E tax credit and one that will potentially have a greater effect on research rather than development spending.

8 RESEARCH COLLABORATIONS

$\mathcal{R}$esearch collaborations represent a public/private partnership not only because both public and private organization participate in them, but also because public resources are used to encourage their formation.[1]

Specific research collaborations are discussed in subsequent chapters. Here, a brief history of research partnerships in the United States is given as an introductory overview.

SEMICONDUCTOR RESEARCH CORPORATION

One of the first formal research collaborations in the United States was the Semiconductor Research Corporation (SRC). A brief history of the SRC will serve to illustrate that many research collaborations or partnerships are formed to address industry-wide technological issues, or at least issues that affect a sizeable segment of the industry. This brief history is also interesting because it illustrates, among other things, a purposeful entrepreneurial response to competitive market conditions.

In the late 1950s, an integrated circuit (IC) industry emerged in the United States. The fledgling industry took form in the 1960s and experienced rapid growth throughout the 1970s. In 1979, when Japanese companies captured 42 percent of the U.S. market for 16 kbit DRAMs (memory devices) and converted Japan's integrated circuit trade balance with the United States from a negative $122 million in 1979 to a positive

---

[1] This chapter is based on Link, Teece, and Finan (1996), Link (1999b), and much of the the underlying research by Link that was sponsored by the National Science Foundation. Link (1999b) was later expanded in Audretsch et al. (2002) and then reproduced in book form in Feldman, Link, and Siegel (2002).

$40 million in 1980, the U.S. industry became painfully aware that its dominance of the IC industry was being seriously challenged. It was clear to all in the industry that it was in their collective best interest to invest in an organizational structure that would strengthen the industry's position in the global semiconductor marketplace.

The Semiconductor Industry Association (SIA) was formed in 1977 to collect and assemble reliable information on the industry and to develop mechanisms for addressing industry issues with the federal government. In a presentation at an SIA Board Meeting in June 1981, Erich Bloch of IBM described to the industry the nature of the growing competition with Japan and proposed the creation of a "semi-conductor research cooperative" to assure continued U.S. technology leadership. This event witnessed the birth of the SRC. In December 1981, Robert Noyce, then SIA chairman and vice-chairman of Intel, announced the establishment of the SRC for the purpose of stimulating joint research in advanced semiconductor technology by industry and U.S. universities and to reverse the declining trend in semiconductor research investments. The SRC was formally incorporated in February 1982 with a stated purpose to:[2]

- Provide a clearer view of technology needs.
- Fund research to address technology needs.
- Focus attention on competition.
- Reduce research redundancy.

Policy makers soon noticed the virtues of cooperative research in part because such organizational structures had worked well in Japan and in part because the organizational success of the SRC demonstrated that cooperation among competitive firms at the fundamental research level was feasible.

SEMATECH

In 1986 when the Semiconductor Industry Association (SIA) and the Semiconductor Research Corporation (SRC) began to explore the possibility of joint industry/government cooperation, the U.S.

---

[2] The eleven founding members were Advanced Micro Devices, Control Data Corporation, Digital Equipment Corporation, General Instrument, Honeywell, Hewlett-Packard, IBM, Intel, Monolithic Memories, Motorola, National Semiconductor, and Silicon Systems.

semiconductor industry was not in a favorable economic position. During 1986, Japan overtook the United States for the first time in terms of their share of the world semiconductor market. Japan had about 45 percent of the world market compared to about 42 percent for the United States. The U.S. semiconductor industry expected Japan's share to grow at the expense of that of the United States.

In January 1987, President Reagan recommended $50 million in matching federal funding for R&D related to semiconductor manufacturing, and this was to be part of the Department of Defense's 1988 budget. Soon thereafter, the SIA approved the formation of SEMATECH and the construction of a world-class research facility.[3] In September 1987, Congress authorized $100 million in matching funding for SEMATECH.

SEMATECH and its members have a mission to:

> ... create a shared competitive advantage by working together to achieve and strengthen manufacturing technology leadership.

This shared vision is accomplished by joint sponsorship of leading edge technology development in equipment supplier companies. As these companies become world class manufacturers, so will the members of SEMATECH.

By 1988, Japan's world market share reached over 50 percent, and that of the United States fell to about 37 percent. The U.S. share declined again in 1989 and then it began to increase at the expense of that of Japan. Early in 1992, the United States was again at parity with Japan at about 42 percent, and stayed slightly ahead of Japan until 1995 when the gap began to widen.

The mid-1990s saw increasing cooperation between U.S. and Japanese semiconductor companies, and in fact, in 1998 International SEMATECH began operations with Hyundai (Japan) and Philips (Amsterdam) as important members. Also, beginning in 1998, all funding came only from member companies.

---

[3] The thirteen charter members of SEMATECH were: Advanced Micro Devices, AT&T, Digital Equipment Corporation, Harris Corporation, Hewlett-Packard Company, IBM Corporation, Intel Corporation, LSI Logic Corporation, Micron Technology, Inc., Motorola, Inc., National Semiconductor Corporation, Rockwell International Corporation, and Texas Instruments, Inc.

Table 8.1 shows selected public/private partnership legislation to encourage research collaborations. Selected examples of research collaborations, based on these legislative initiatives, are discussed in subsequent chapters.

Table 8.1.   Selected Public/Private Partnership Legislation to Encourage
Research Collaboration

Enabling Legislation	Characteristics of the Program
Stevenson-Wylder Technology Innovation Act of 1980	Act predicated on the premise that federal laboratories embody industrially-useful technology. Federal laboratory mandated to establish an Office of Research and Technology Application to facilitate transfer of public technology to the private sector.
University and Small Business Patent Procedure Act of 1980	Known as Bayh-Dole Act. Reformed federal patent policy by providing increased incentives for diffusion of federally-funded innovation results. Universities, non-profit organizations, and small businesses permitted to obtain titles to innovations developed with governmental funds; federal agencies to grant exclusive licenses to their technology to industry.
Small Business Innovation Development Act of 1982	Act required federal agencies to provide special funds to support small business R&D that complemented the agency's mission. Programs called Small Business Innovation Research (SBIR) programs. Act reauthorized in 1992.
National Cooperative Research Act of 1984	NCRA encouraged formation of joint research venture among U.S. firms. Amended by the National Cooperative Research and Production Act of 1993, thereby expanding antitrust protection to joint production ventures.
Trademark Clarification Act of 1984	Act set forth new licensing and royalty regulations to take technology from federally-funded facilities into the private sector. Specifically permitted government-owned, contractor-operated (GOCO) laboratories to make decisions regarding which patents to license to the private sector, and contractors could receive royalties on such patents.
Federal Technology Transfer Act of 1986	Act amended by the Stevenson-Wylder Act. Made technology transfer an explicit responsibility of all federal laboratory scientists and engineers. Authorized cooperative research and development agreements (CRADAs). Amended by the National Competitiveness Technology Transfer Act of 1989 to include contractor operated laboratories.

Omnibus Trade and Competitiveness Act of 1988	Act established Advanced Technology Program (ATP) and Manufacturing Extension Partnership (MEP) within the re-named National Institute of Standards and Technology (NIST).
Defense Conversion, Reinvestment, and Transition Assistance Act of 1992	Act created infrastructure for dual-use partnerships. Through Technology Reinvestment Project partnerships, Department of Defense given the ability to leverage the potential advantages of advanced commercial technologies to meet departmental needs.
National Technology Transfer and Advancement Act of 1995	Act gave CRADA partners sufficient intellectual property rights to justify prompt commercialization of inventions resulting from CRADAs.
Technology Transfer Commercialization Act of 2000	Act improved the ability of federal agencies to license federally-owned inventions by reforming technology training authorities under the Bayh-Dole Act.

9 RESEARCH JOINT VENTURES

*R*esearch joint ventures (RJVs) represent a public/private partnership.[1] An RJV is a collaborative research arrangement through which firms jointly acquire technical knowledge.

In terms of the taxonomy used to characterize public/private partnerships, RJVs, and the legislation that promulgates them, represent indirect governmental involvement in innovation. The economic objective of tax incentives, in particular the R&E tax credit, is to leverage private R&D. See Table 9.1.

PUBLIC POLICY TOWARD RESEARCH JOINT VENTURES

To place the activities surrounding the SRC's formation (Chapter 8) in a broader context, there was growing concern in the late 1970s and early 1980s regarding the pervasive slowdown in productivity growth throughout the U.S. industrial sector. More specifically, policy makers were troubled by the declining global market shares of leading American companies, especially firms in the semiconductor industry.[2]

---

[1] This chapter is based on Link and Bauer (1989), Link (1999b), and Link and Scott (2005b).

[2] The declining U.S. position in the semiconductor industry was well known and in other industries there was widespread concern although the empirical evidence about the competitive position of the United States in international markets was incomplete. However, when the U.S. Department of Commerce (1990) released its 1990 report on emerging technologies, it was apparent to all that the concerns expressed in the early 1980s were quite valid.

Table 9.1. Taxonomy of Public/Private Partnerships

	Economic Objective	
Governmental Involvement	*Leverage Public R&D*	*Leverage Private R&D*
Indirect		Patent system
		(Patent Act)
		Tax incentives
		(R&E tax credit)
		Research joint ventures
		(NCRA and NCRPA)
Direct		
Financial Resources		
Infrastructural Resources		
Research Resources		

As noted in a November 18, 1983 House report about the proposed Research and Development Joint Ventures Act of 1983:

> A number of indicators strongly suggest that the position of world technology leadership once firmly held by the United States is declining. The United States, only a decade ago, with only five percent of the world's population was generating about 75 percent of the world's technology. Now, the U.S. share has declined to about 50 percent and in another ten years, without fundamental changes in our Nation's technological policy ... the past trend would suggest that it may be down to only 30 percent. [In hearings,] many distinguished scientific and industry panels had recommended the need for some relaxation of current antitrust laws to encourage the formation of R&D joint ventures. ... The encouragement and fostering of joint research and development ventures are needed responses to the problem of declining U.S. productivity and international competitiveness. According to the testimony received during the Committee hearings, this legislation will provide for a significant increase in

the efficiency associated with firms doing similar research and development and will also provide for more effective use of scarce technically trained personnel in the United States.

In an April 6, 1984 House report on competing legislation, the Joint Research and Development Act of 1984, the supposed benefits—and recall that at this time it was still too soon for there to be visible benefits coming from the SRC's activities on behalf of the IC industry—of joint research and development were for the first time clearly articulated:

Joint research and development, as our foreign competitors have learned, can be procompetitive. It can reduce duplication, promote the efficient use of scarce technical personnel, and help to achieve desirable economies of scale. ... [W]e must ensure to our U.S. industries the same economic opportunities as our competitors, to engage in joint research and development, if we are to compete in the world market and retain jobs in this country.

The National Cooperative Research Act (NCRA) of 1984, after additional revisions in the initiating legislation, was passed on October 11, 1984:

... to promote research and development, encourage innovation, stimulate trade, and make necessary and appropriate modifications in the operation of the antitrust laws.

The NCRA of 1984 created a registration process, later expanded by the National Cooperative Research and Production Act (NCRPA) of 1993 and the Standards Development Organization Advancement Act of 2004 (SDOAA), under which RJVs can voluntarily disclose their research intentions to the U.S. Department of Justice; all disclosures are made public in the *Federal Register*.

RJVs gain two significant benefits from filing with the Department of Justice. One, if the venture were subjected to criminal or civil antitrust action, the courts would evaluate the alleged anticompetitive behavior under a rule of reason rather than presumptively ruling that the behavior constituted a *per se* violation of the antitrust law. For RJVs that have filed, the Act states:

> In any action under the antitrust laws ... the conduct of any person in making or performing a contract to carry out a joint research and development venture shall not be deemed illegal per se; such conduct shall be judged on the basis of its reasonableness, taking into account all relevant factors affecting competition, including, but not limited to, effects on competition in properly defined, relevant research and development markets.

And two, if the venture were found to fail a rule-of-reason analysis it would be subject to actual damages rather than treble damages.

One of the more notable RJVs formed and made public through the NCRA disclosure process was SEMATECH (Chapter 8). It was thought that SEMATECH would be the U.S. semiconductor industry's/U.S. government's response to the Japanese government's targeting of their semiconductor industry for global domination.

TRENDS IN RJVs

Through 2003, there have been 913 formal RJVs filed under the NCRA. Certainly, this number is a lower bound on the total number of research partnerships in the United States, even since 1984. Not all RJVs are as publicly visible as SEMATECH. Most are quite small, with only two or three members, and others are quite large with hundreds of members. The average size of one of these joint ventures is about 13 members.[3]

As an illustration of the research activity that can successfully occur through a small, less visible research partnership, consider the Southwest Research Institute Clean Heavy Diesel Engine II joint venture, noticed in the *Federal Register* in early-1996. The eleven member companies, from six countries including the United States, joined together to solve a common set of technical problems. Diesel engine manufacturers were having difficulties, on their own, meeting desired emission control levels. The eleven companies were coordinated by Southwest Research Institute, an independent, non-profit contract research organization in San Antonio,

---

[3] The range of membership size is large; 2 to 539.

Texas, to collaborate on the reduction of exhaust emissions. The joint research was successful, and each member company took with it fundamental process technology to use in their individual manufacturing facilities to meet desired emission control levels. The joint venture was formally disbanded in mid-1999.

Figure 9.1 shows the trend in RJVs from 1985 through 2003 based on the year of disclosure in the *Federal Register*.

Figure 9.1. Number of RJVs Disclosed in the *Federal Register*, by Year of Disclosure

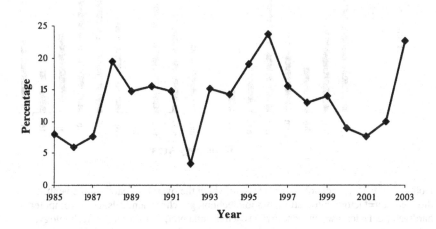

Certainly, the trend in RJV disclosures was upward until the mid-1990s, and since then it has generally declined until 2002.[4] While informal cooperation in research may have been prevalent in the United States for decades, formal RJV relationships are new and it will take longer than a decade and a half to detect meaningful trends.

---

[4] Brod and Link (2001) have identified selected correlates with the trend in RJVs over time. In particular, the annual number of filings of RJVs changes, on average, in a countercyclical manner and in relationship to industrial development (as opposed to research) activity.

Figure 9.2 shows the percent of RJVs disclosed in the *Federal Register*, by technology area. Collaboration occurs most often when the research area is computer software or a defense-related area.

Figure 9.2. Percentage of RJVs Disclosed in the *Federal Register*, by
           Technology Area

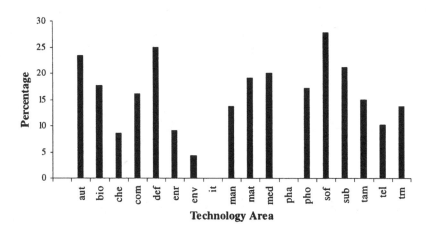

Note: The primary technology area toward which the overall research of the venture is directed: aut=factory automation; bio=biotechnology; che=chemicals; com=computer hardware; def=defense; enr=energy; env=environmental; it=information technology; man=manufacturing equipment; mat=advanced materials; med=medicals; pha=pharmaceuticals; pho=photonics; sof=computer software; sub=subassemblies and components; tam=test and measurement; tel=telecommunications; trn=transportation.

Although RJVs as formal entities are relatively new to the technology strategy arena, the literature concludes that there are both benefits and costs to members of the venture. Following Hagedoorn, Link, and Vonortas (2000), the benefits include:

- The opportunity for participants to capture knowledge spillovers from other members.
- Reduced research costs due to a reduction in duplicative research.
- Faster commercialization since the fundamental research stage is shortened.

- The opportunity to develop industry-wide competitive vision.

The costs include:

- A lack of appropriability since research results are shared among the participants.
- Managerial tension, in some cases, as participants learn to trust each other and to work together.

Research partnerships are correctly viewed as a complementary source of technical knowledge and technical efficiency for the firm. Thus, firms that participate in a research partnership leverage their own R&D process through interactions and knowledge sharing.

RJV PARTNERS

Especially noticeable in the RJVs filed with the Department of Justice is the presence of universities as research partners.[5] Over the past 18 years, the number of RJVs with at least one university partner has increased. On average, nearly 15 percent of RJVs have, as of 2003, at least one university research partner, and of these over 90 percent are U.S. universities. Those RJVs with universities as research partners have, on average, 5 university partners.

A university has a financial incentive to partner with industry in its applied research, especially if commercial technologies are expected to

_____

[5] The federal government also enters directly into research partnerships with firms through the federal laboratory system. This relationship can take various forms ranging from informal relationships whereby a firm(s) interacts with a federal laboratory scientist, or more formal relationships whereby a firm(s) utilizes federal laboratory facilities and is jointly involved with the laboratory scientists in the research. Or, the relationship can be nothing more than an informational transfer whereby the firm utilizes public information that was generated within a government agency. While very few studies have systematically looked at the economics of federal laboratories as research partners, two generalizations can be made (Leyden and Link 1999):

- Federal laboratories are generally associated with research joint ventures that are large in terms of other member companies.
- One key advantage to partnering with a federal laboratory is access to specialized technical equipment.

result. Industry has a research efficiency incentive to partner with a university. Efficiencies are gained through access to complementary activities and research results, and access to key university personnel (faculty and graduating students).

As Rosenberg and Nelson (1994, p. 340) note:

> What university research most often does … is to stimulate and enhance the power of R&D done in industry, as contrasted with providing a substitute for it.

Relatedly, Hall, Link, and Scott (2003, p. 490) argue:

> Universities are included (invited by industry) in those research projects that involve what we have called "new" science. Industrial research participants perceive that the university could provide research insight that is anticipatory of future research problems and could be an ombudsman anticipating and translating to all the complex nature of the research being undertaken. Thus, one finds universities purposively involved in projects that are characterized as problematic with regard to the use of basic knowledge.

Generalizations aside, some stylized conclusions can be drawn from the limited empirical investigations (Hall, Link, Scott 2000, 2001; Link and Scott 2005b):

- Firms that interact with universities generally have greater R&D productivity and greater patenting activity.
- One key motive for firms to maintain joint research relationships with universities is to have access to key university personnel—faculty as well as students as potential employees.
- Larger RJVs are more likely to invite a university to join the venture than smaller RJVs because larger RJVs are less likely to expect substantial additional appropriability problems to result because of the addition of a university partner, and because the larger ventures have both a lower marginal cost and a higher marginal value from the university R&D contributions to the venture's innovative output.

Many commentators predict that university participation in such collaboration will increase in the future. According to the Council on Competitiveness (1996, pp. 3-4):

> Over the next several years, participation in the U.S. R&D enterprise will have to continue experimenting with different types of partnerships to respond to the economic constraints, competitive pressures and technological demands that are forcing adjustment across the board. ... [and in response] industry is increasingly relying on partnerships with universities, while the focus of these partnerships is shifting progressively toward involvement in shorter-term research.

And (Council on Competitiveness 1996, p. 11):

> For universities, cutbacks in defense spending have resulted in a *de facto* reallocation of funding away from the physical sciences and engineering and shifted the focus of defense research away from the frontiers of knowledge [e.g., basic science] to more applied efforts. ... Although defense spending is clearly not the only viable mechanism to support frontier research and advanced technology, the United States has yet to find an alternative innovation paradigm to replace it.

Given this spending trend, and the increasing ease of global technology transfer, it is conceivable, at least according to the Council, that the United States may lose its technological leadership in some important areas such as health and advanced materials, since innovation in these fields is closely linked to improvements in basic science.

There is some indication that scholars are beginning to think more deeply and more broadly about the social, economic, and technological consequences of university involvement in private sector research partnerships (Siegel et al. 2001). This thinking reflects some major concerns about the impact of these relationships on the research university's mission to conduct basic research. Unfortunately, there is a void of information that can be studied by researchers to examine the ramifications of this trend from a wide variety of disciplinary perspectives.

It is likely that the increasing trend toward university private-sector research partnerships will continue. A 1993 national survey of U.S.

biology, chemistry, and physics faculty members revealed that many academic scientists desired more of such involvement. An earlier survey of engineering faculty members reached the same conclusion. However, one of the authors (Morgan 1998, p. 169) of the surveys was quick to point out one area of major concern:

> ... a diminution of the role of the university as an independent voice to help look out for the broader societal good and to guard against industrial as well as other excesses. An independent science, engineering and public policy role is essential to ensure an adequate supply of well educated scientists and engineers prepared to work with the public sector in public interest groups. Having industry assume a more central role as customer and client for university-based scientific and engineering research, while in some way a natural and desirable step, needs to be balanced against the need for independence, oversight and service to society and the larger public good.

UNIVERSITY-BASED RESEARCH PARKS

University-based research parks are a form of a public/private partnership between a university and tenant firms.[6] While not a formal RJV, parks represent an infrastructure conducive to the formation of new RJVs or an environment conducive for universities to become a partner in an existing RJV.

A number of definitions of a research or science park have been proffered by various institutions or associations. Following Link and Scott (forthcoming c):

> A university research park is a cluster of technology-based organizations that locate on or near a university campus in order to benefit from the university's knowledge base and ongoing research. The university not only transfers knowledge but expects to develop knowledge more effectively given the association with the tenants in the research park.

---

[6] The term science park is used more commonly in European Union countries. This section draws on Link and Scott (forthcoming c).

Generally, if the park is on or adjacent to a university campus the university owns the park land and either oversees, or at least advises on, aspects of the activities that take place in the park as well as on the strategic direction of the park's growth.[7, 8] When the park is located off campus, it is often the case that the park land is owned by a private venture—and sold or leased to tenants—but the university had contributed financial capital to its formation and/or intellectual capital to its operation; therefore, there are elements of an administrative relationship between the university and these research parks.[9]

Universities are motivated to develop a research park on their own or in partnership by the possibility of financial gain associated with technology transfer, the opportunity to have faculty and students interact at the applied level with technology-based organizations, and by the responsibility of contributing to a regional-based economic development effort.[10, 11] Research organizations are motivated to locate in a research park to gain access to faculty, students, and research equipment, and to foster research synergies.

Based on the definition above, the population of currently active university-based research parks is shown in Figure 9.3. Notable in the figure are the following parks: Stanford Research Park (established in 1951), Cornell Business & Technology Park (established in 1952); and the

---

[7] Such oversight may include tenant criteria for leasing space in the park (Link and Link 2003). Such criteria may specify particular technologies or state that the tenant must maintain an active research relationship with university departments and their students.

[8] Approximately 6 percent of existing parks are formally affiliated with more than one university (e.g., Duke University, North Carolina State University, and University of North Carolina have a formal relationship with Research Triangle Park.).

[9] The form of the relationship between the university and the research park can be very explicit, as in the case when the university owns the park land and buildings and leases space to criteria-specific tenants; or very implicit, as in the case when the privately-owned park is juxtaposed to the university and the university owns and operates buildings on park land. Certainly, a physical relationship between the university and the park does not necessarily imply an administrative or strategic relationship.

[10] In most cases, regional economic development is one justification of the creation of a university-related research park.

[11] Just over 50 percent of university-related research parks were initially funded with public moneys. Of those, the public sector supported about 70 percent of the initial park cost.

Research Triangle Park of North Carolina (established in 1959). Also notable in the figure is the increase in park formation that began in the late-1970s and accelerated in the early 1980s.[12]

Figure 9.3. University-based Research Parks by Year Formed, 1951-2004

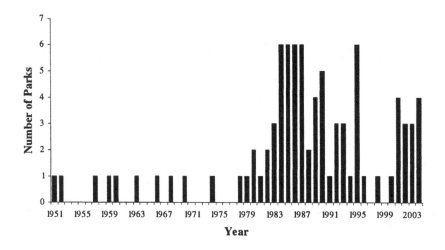

During the early and mid-1970s, real industrial R&D spending decreased. It was not until 1977 that real R&D performed in industry was able to return to its 1969-pre-decline level, and relatedly, in 1978 park formations began to increase. It is reasonable to hypothesize that private sector demand for research park space increased during this R&D growth period because firms were looking for cooperative research partnerships to expand their research portfolios, as opposed to development portfolios.

The period of the relatively rapid increase in park formation corresponds to a period of significant public policy initiatives to encourage university-with-industry relationships, increases in industrial R&D spending, and the formation of cooperative research partnerships. The Bayh-Dole Act was passed in 1980 (Chapter 8), the R&E tax credit was enacted in 1981 (Chapter 7), and the National Cooperative Research Act

---

[12] Danilov (1971) attributes the relatively long period from about 1960 to the early 1970s, during which the research park movement seemingly stalled, to a number of park efforts that failed as well as to the restraints on corporate R&D growth because of a lackluster economy.

was legislated in 1984 (discussed above). All of these public initiatives leveraged private sector R&D activity, which could have stimulated states and universities to establish potentially beneficial locations for their R&D to take place.

10 ADVANCED TECHNOLOGY PROGRAM

*T*he Advanced Technology Program (ATP) is a public/private partnership, and, as shown in Table 10.1, it leverages private R&D through direct involvement through the provision of financial resources.[1]

Table 10.1. Taxonomy of Public/Private Partnerships

	Economic Objective	
Governmental Involvement	*Leverage Public R&D*	*Leverage Private R&D*
Indirect		Patent system (Patent Act)
		Tax incentives (R&E tax credit)
		Research joint ventures (NCRA and NCRPA)
Direct		
Financial Resources		**Advanced Technology Program** (Omnibus Trade and Competitiveness Act)
Infrastructural Resources		
Research Resources		

---

[1] This chapter draws on Link (1996b), Link and Scott (1998), and Link (1999b).

The ATP was established within the National Institute of Standards and Technology (NIST, see Chapter 11) through the Omnibus Trade and Competitiveness Act of 1988, and modified by the American Technology Preeminence Act of 1991. The goals of the ATP, as stated in its enabling legislation, are to assist U.S. businesses in creating and applying the generic technology and research results necessary to:

> (1) commercialize significant new scientific discoveries
>   and technologies rapidly
> (2) refine manufacturing technologies.

These same goals were restated in the *Federal Register* on July 24, 1990:

> The ATP . . . will assist U.S. businesses to improve their
> competitive position and promote U.S. economic growth
> by accelerating the development of a variety of pre-
> competitive generic technologies by means of grants and
> cooperative agreements.

As shown in Figure 4.1, TFP significantly declined in the early 1980s. While it recovered in the mid-1980s, the U.S. position in critical technologies, relative to Japan and to Europe, was not strong. Table 10.2 shows the technologies where, in 1989, the U.S. was still behind, and Table 10.3 shows the technologies where the trend remained unfavorable. In general, computer-based operating and processing technologies were, at that time, no longer a strength of the United States.

The ATP received its first appropriation from Congress in FY 1990. The program funds research, not product development. Commercialization of the technology resulting from a project might overlap the research effort at a nascent level, but generally full translation of the technology into products and processes may take a number of additional years. ATP, through cost sharing with industry, invests in risky technologies that have the potential for spillover benefits to the economy.

Appropriations to ATP increased from $10 million in 1990 to a peak of $341 million in 1995. Funding decreased in 1996 to $221 million, and it has averaged about $200 million per year until 2004 when it fell to just under $150 million. To date, ATP has funded through competitive processes approximately 770 research projects involving over 1,500 organizations. In total, ATP has awarded over $2.0 billion with industry allocating nearly that same amount in the form of research matching funds.

Table 10.2. Critical Technology Report Card, Status in 1989

Relative Position.	U.S. vs Japan	U.S. vs Europe
Behind	Advanced Materials Advanced Semiconductor Devices High-Density Data Storage Optoelectronics	Digital Imaging Technology
Even	Superconductors	Flexible Computer-Integrated Manufacturing Superconductors
Ahead	Artificial Intelligence Biotechnology Flexible Computer-Integrated Manufacturing High-Performance Computing Medical Devices and Diagnostics Sensor Technology	Artificial Materials Advanced Semiconductor Devices Artificial Intelligence Biotechnology High-Density Data Storage High-Performance Computing Medical Devices and Diagnostics Optoelectronics Sensor Technology

Source: Department of Commerce (1990).

ATP has provided incentives to firms to undertake research that would not otherwise have been pursued—like projects A or B in Figure 3.1 (Link and Scott 2001). As an illustration of the research that ATP has funded, consider the following.

A printed wiring board (PWB) or printed circuit board (PCB) is a device that provides electrical interconnections and a surface for mounting electrical components. The United Stated dominated the world PWB market in the early 1980s, enjoying a 42 percent world market share by 1984. In 1985, that share fell to 36 percent against Japan's increasing market share. By 1987, Japan's world market share surpassed that of the United States for the first time (30 percent compared to 29 percent), and

Japan's share eventually peaked at 35 percent in 1990, compared to the U.S. share of 26 percent.

Table 10.3. Critical Technology Report Card, Trends in 1989

Relative Trend	U.S. vs Japan	U.S. vs Europe
Losing Badly	Advanced Materials Biotechnology Digital Imaging Technology Superconductors	Digital Imaging Technology Flexible Computer- Integrated Manufacturing
Losing	Advanced Semiconductor Devices High-Density Data Storage High-Performance Computing Medical Devices and Diagnostics Optoelectronics Sensor Technology	Medical Devices and Diagnostics
Holding	Artificial Intelligence Flexible Computer- Integrated Manufacturing	Advanced Materials Advanced Semiconductor Devices High-Density Data Storage Optoelectronics Sensor Technology Superconductors
Gaining		Artificial Intelligence Biotechnology High-Performance Computing

Source: Department of Commerce (1990).

In 1991, the Council on Competitiveness released a report documenting that the U.S. PWB industry had lost its global competitive advantage. In April 1991, ATP announced an award to a joint venture led by the National Center for Manufacturing Sciences (NCMS) to research aspects of PWB interconnect systems. ATP contributed nearly $14 million to the project, matched by firms in the venture. The joint venture had a number of technical successes and, according to the members of the joint venture, these ATP leveraged technical successes helped to increase the competitive position of the industry in the world market.

11 NATIONAL INSTITUTE OF STANDARDS AND TECHNOLOGY

The National Institute of Standards and Technology (NIST) is a public/private partnership, and as shown in Table 11.1, it provides direct infrastructural and research resources to leverage both public and private R&D.[1]

NATIONAL INSTITUTE OF STANDARDS AND TECHNOLOGY

A standard is a prescribed set of rules, conditions, or requirements concerning:

- Definitions of terms.
- Classification of components.
- Specification of materials, their performance, and their operations.
- Delineation of procedures.
- Measurement of quantity and quality in describing materials, products, systems, services, or practices.

To understand the current activities that take place at NIST, its public good mission must be placed in an historical perspective. The concept of the government's involvement in standards traces to the Articles of Confederation signed on July 9, 1778. In Article 9, § 4:

> The United States, in Congress assembled, shall also have
> the sole and exclusive right and power of regulating the

---

[1] This chapter draws on National Research Council (1995), Link (1996b, 1999b), and Link and Scott (1998b).

alloy and value of coin struck by their own authority, or by that of the respective States; fixing the standard of weights and measures throughout the United States ...

Table 11.1. Taxonomy of Public/Private Partnerships

Governmental Involvement	Economic Objective	
	Leverage Public R&D	*Leverage Private R&D*
Indirect		Patent system (Patent Act)
		Tax incentives (R&E tax credit)
		Research joint ventures (NCRA and NCRPA)
Direct		
Financial Resources		Advanced Technology Program (Omnibus Trade and Competitiveness Act)
Infrastructural Resources	**National Institute of Standards and Technology** (Organic Act)	**National Institute of Standards and Technology** (Organic Act)
Research Resources	**National Institute of Standards and Technology** (Organic Act)	**National Institute of Standards and Technology** (Organic Act)

This responsibility was reiterated in Article 1, § 8 of the Constitution of the United States:

The Congress shall have power ... To coin money, regulate the value thereof, and of foreign coin, and fix the standard of weights and measures ...

On July 20, 1866, Congress and President Andrew Johnson authorized the use of the metric system in the United States. This was formalized in the Act of 28 July 1866—An Act to Authorize the Use of the Metric System of Weights and Measures:

> *Be it enacted* ..., That from and after the passage of this act it shall be lawful throughout the United States of America to employ the weights and measures of the metric system; and no contract or dealing, or pleading in any court, shall be deemed invalid or liable to objection because the weights or measures expressed or referred to therein are weights and measures of the metric system.

As background to this Act, the origins of the metric system can be traced to the research of Gabriel Mouton, a French vicar, in the late 1600s. His standard unit was based on the length of an arc of 1 minute of a great circle of the earth. Given the controversy of the day over this measurement, the National Assembly of France decreed on May 8, 1790 that the French Academy of Sciences along with the Royal Society of London deduced an invariable standard for all the measures and all the weights.

Within a year, a standardized measurement plan was adopted based on terrestrial arcs, and the term mètre (meter), from the Greek *metron* meaning to measure, was assigned by the Academy of Sciences (Link 1996b).

Because of the growing use of the metric system in scientific work rather than commercial activity, the French government held an international conference in 1872, which included the participation of the United States, to settle on procedures for the preparation of prototype metric standards. Then, on May 20, 1875, the United States participated in the Convention of the Meter in Paris and was one of the eighteen signatory nations to the Treaty of the Meter.

In a Joint Resolution before Congress on March 3, 1881, it was resolved that:

> The Secretary of the Treasury be, and he is hereby directed to cause a complete set of all the weights and measures adopted as standards to be delivered to the governor of each State in the Union, for the use of agricultural colleges in the States, respectively, which have received a grant of lands from the United States, and

also one set of the same for the use of the Smithsonian Institution.

Then, the Act of 11 July 1890 gave authority to the Office of Construction of Standard Weights and Measures (or Office of Standard Weights and Measures), which had been established in 1836 within the Treasury's Coast and Geodetic Survey:

> For construction and verification of standard weights and measures, including metric standards, for the custom-houses, and other offices of the United States, and for the several States ...

The Act of 12 July 1894 established standard units of electrical measure:

> *Be it enacted* ..., That from and after the passage of this Act the legal units of electrical measure in the United States shall be as follows: ... That it shall be the duty of the National Academy of Sciences [established in 1863] to prescribe and publish, as soon as possible after the passage of this Act, such specifications of detail as shall be necessary for the practical application of the definitions of the ampere and volt hereinbefore given, and such specifications shall be the standard specifications herein mentioned.

Following from a long history of our Nation's leaders calling for uniformity in science, traceable at least to the several formal proposals for a Department of Science in the early 1880s, and coupled with the growing inability of the Office of Weights and Measures to handle the explosion of arbitrary standards in all aspects of federal and state activity, it was inevitable that a standards laboratory would need to be established. The political force for this laboratory came in 1900 through Lyman Gage, then Secretary of the Treasury under President William McKinley.

Gage's original plan was for the Office of Standard Weights and Measures to be recognized as a separate agency called the National Standardizing Bureau. This Bureau would maintain custody of standards, compare standards, construct standards, test standards, and resolve problems in connection with standards. Although Congress at that time wrestled with the level of funding for such a laboratory, the importance of the laboratory was not debated.

Finally, the Act of 3 March 1901, also known as the Organic Act, established the National Bureau of Standards within the Department of the Treasury, where the Office of Standard Weights and Measures was administratively located:

> *Be it enacted by the Senate and House of Representatives of the United States of America in Congress assembled,* That the Office of Standard Weights and Measures shall hereafter be known as the National Bureau of Standards ... That the functions of the bureau shall consist in the custody of the standards; the comparison of the standards used in scientific investigations, engineering, manufacturing, commerce, and educational institutions with the standards adopted or recognized by the Government; the construction, when necessary, of standards, their multiples and subdivisions; the testing and calibration of standard measuring apparatus; the solution of problems which arise in connection with standards; the determination of physical constants and the properties of materials, when such data are of great importance to scientific or manufacturing interests and are not to be obtained of sufficient accuracy elsewhere.

The Act of 14 February 1903 established the Department of Commerce and Labor, and in that Act it was stated that the National Bureau of Standards be moved from the Department of the Treasury to the Department of Commerce and Labor.

Then, in 1913, when the Department of Labor was established as a separate entity, the Bureau was formally housed in the Department of Commerce.

In the post World War I years, the Bureau's research focused on assisting in the growth of industry. Research was conducted on ways to increase the operating efficiency of automobile and aircraft engines, electrical batteries, and gas appliances. Also, work was begun on improving methods for measuring electrical losses in response to public utility needs. This latter research was not independent of international efforts to establish electrical standards similar to those established over 50 years earlier for weights and measures.

After World War II, significant attention and resources were given to the activities of the Bureau. In particular, the Act of 21 July 1950 established standards for electrical and photometric measurements:

> *Be it enacted by the Senate and House of Representatives of the United States of America in Congress assembled,* That from and after the date this Act is approved, the legal units of electrical and photometric measurements in the United States of America shall be those defined and established as provided in the following sections. ...
> The unit of electrical resistance shall be the ohm. ...
> The unit of electrical current shall be the ampere. ...
> The unit of electromotive force and of electrical potential shall be the volt. ...
> The unit of electrical quantity shall be the coulomb. ...
> The unit of electrical capacity shall be the farad. ...
> The unit of electrical inductance shall be the henry. ...
> The unit of power shall be the watt. ...
> The units of energy shall be the (a) joule ... and (b) the kilowatt-hour. ...
> The unit of intensity shall be the candle. ...
> The unit of flux light shall be the lumen. ...
> It shall be the duty of the Secretary of Commerce to establish the values of the primary electric and photometric units in absolute measure, and the legal values for these units shall be those represented by, or derived from, national reference standards maintained by the Department of Commerce.

Then, as a part of the Act of 20 June 1956, the Bureau moved from Washington, DC to Gaithersburg, Maryland. The responsibilities listed in the Act of 21 July 1950, and many others, were transferred to the National Institute of Standards and Technology (NIST) when the National Bureau of Standards was renamed under the guidelines of the Omnibus Trade and Competitiveness Act of 1988:

> The National Institute of Standards and Technology [shall] enhance the competitiveness of American industry while maintaining its traditional function as lead national laboratory for providing the measurement, calibrations, and quality assurance techniques which underpin United

> States commerce, technological progress, improved product reliability and manufacturing processes, and public safety ... [and it shall] advance, through cooperative efforts among industries, universities, and government laboratories, promising research and development projects, which can be optimized by the private sector for commercial and industrial applications ... [More specifically, NIST is to] prepare, certify, and sell standard reference materials for use in ensuring the accuracy of chemical analyses and measurements of physical and other properties of materials ...

NIST's mission is to promote U.S. economic growth by working with industry to develop and apply technology, measurements, and standards. As a group, these represent infrastructure technology or infratechnologies (Chapter 5). NIST carries out this mission through four major programs including ATP, but its centerpiece program is the measurement and standards laboratories program. It provides technical leadership for vital components of the nation's technology infrastructure needed by U.S. industry to continually improve its products and services.

NIST's organizational structure is laboratory based. The laboratories at NIST provide technical leadership for vital components of the Nation's technology infrastructure needed by U.S. industry to continually improve its products and services. Currently, there are seven research laboratories at NIST:[2]

- The *Electronics and Electrical Engineering Laboratory* (EEEL) promotes U.S. economic growth by providing measurement capability of high impact focused primarily on the critical needs of the U.S. electronics and electrical industries, and their customers and suppliers.
- The *Manufacturing Engineering Laboratory* (MEL) performs research and development of measurements, standards, and infrastructure technology as related to manufacturing.
- The *Chemical Science and Technology Laboratory* (CSTL) provides chemical measurement infrastructure to enhance U.S. industry's

---

[2] In addition to these research laboratories, *Technology Services* provides a variety of products and services to U.S. industry such as Standard Reference Materials, Standard Reference Data, and Weights and Measures.

productivity and competitiveness; assure equity in trade; and improve public health, safety, and environmental quality.

- The *Physics Laboratory* (PL) supports U.S. industry by providing measurement services and research for electronic, optical, and radiation technologies.
- The *Materials Science and Engineering Laboratory* (MSEL) stimulates the more effective production and use of materials by working with materials suppliers and users to assure the development and implementation of the measurement and standards infrastructure for materials.
- The *Building and Fire Research Laboratory* (BFRL) enhances the competitiveness of U.S. industry and public safety by developing performance prediction methods, measurement technologies, and technical advances needed to assure the life cycle quality and economy of constructed facilities.
- The *Information Technology Laboratory* (ITL) works with industry, research, and government organizations to develop and demonstrate tests, test methods, reference data, proof of concept implementations, and other infrastructural technologies.[3]

As an illustration of research that takes place at NIST, consider the following. Chemical compounds known as chlorofluorocarbons (CFCs) have been used extensively as aerosol propellants, refrigerants, solvents, and industrial form blowing agents. Until the past decade, most refrigerants used throughout the world were made of CFCs because of their desirable physical and economic properties. However, research has shown that the release of CFCs into the atmosphere is possibly damaging the ozone layer of the earth. In response to this research finding, international legislation was drafted that resulted in the signing of the Montreal Protocol in 1987, a global agreement to phase out the production and use of CFCs and replace them with other compounds that would have a lesser impact on the environment. In order to meet the phase-out schedule in the Protocol, research was needed to develop new types of refrigerants, called alternative refrigerants, that would retain the desirable properties of CFCs but would pose little or no threat to the ozone layer. The Protocol called for production and consumption levels to be capped at the 1996 level by year 1990, to then decrease to 80 percent of that level by

---

[3] The Computer Systems Laboratory (CSL) and the Computing and Applied Mathematics Laboratory (CAML) were combined on February 16, 1997 to form the Information Technology Laboratory.

1994, and to decrease again to 50 percent of that level by 1999. In the United States, the 1992 amendment to the Clean Air Act of 1990 called for a faster phase-out schedule in which no CFCs could be produced after 1996. U.S. industries that relied on CFC refrigerants could not meet this schedule on their own, but the CSTL at NIST had been involved in alternative refrigerant research since 1982. That laboratory had developed a tool known as REFPROP, a computer package of standard reference data on *ref*rigerant *prop*erties that could be used by industry to model, in a standardized manner, the behavior of various alternative refrigerant mixtures for the development of CFC replacements. NIST's standard reference data saved the refrigeration industry millions of dollars in redundant research as well as improved environment quality by ensuring that the domestic industry met the phase-out schedule.

THE ECONOMICS OF STANDARDS

An industry standard is a set of specifications to which all elements of products, processes, formats, or procedures under its jurisdiction must conform. The process of standardization is the pursuit of this conformity, with the objective of increasing the efficiency of economic activity.

The complexity of modern technology, especially its system character, has led to an increase in the number and variety of standards that affect a single industry or market. Standards affect the R&D, production, and market penetration stages of economic activity and therefore have a significant collective effect on innovation, productivity, and market structure. Thus, a concern of government policy is the evolutionary path by which a new technology or, more accurately, certain elements of a new technology become standardized.

Standardization, according to Tassey (2000), can and does occur without formal promulgation as a standard. This distinction between *de facto* and promulgated standards is important more from an institutional process than an economic impact perspective.

In one sense, standardization is a form rather than a type of infrastructure because it represents a codification of an element of an industry's technology or simply information relevant to the conduct of economic activity. And, because the selection of one of several available forms of a technology element as the standard has potentially important economic effects, the process of standardization is important.

While economics is increasingly concerned with standards due to their proliferation and pervasiveness in many new high-technology industries, the economic roles of standards are unfortunately poorly understood. Standards can be grouped into two basic categories:

(1) product-element standards, and
(2) nonproduct-element standards.

This distinction is important because the economic role of each type is different.

Product-element standards typically involve one of the key attributes or elements of a product, as opposed to the entire product. In most cases, market dynamics determine product-element standards. Alternative technologies compete intensely until a dominant version gains sufficient market share to become the single *de facto* standard. Market control by one firm can truncate this competitive process. Conversely, nonproduct-element standards tend to be competitively neutral within the context of an industry. This type of standard can impact an entire industry's efficiency and its overall market penetration rate (Link 1983).

Industry organizations often set non-product-element standards using consensus processes. The technical bases (infrastructure technologies or infratechnologies) for these standards have a large public good content. Examples include measurement and test methods, interface standards, and standard reference materials.

From both the positions of a strategically-focused firm as well as a public policy maker, standardization is not an all-or-nothing proposition. In complimented system technologies, such as distributed data processing, telecommunications, or factory automation, standardization typically proceeds in an evolutionary manner in lock step with the evolution of the technology. Complete standardization too early in the technology's life cycle can constrain innovation.

The overall economic value of a standard is determined by its functionality (interaction with other standards at the systems level) and the cost of implementation (compliance costs). Standards should be competitively neutral, which means adaptable to alternative applications of a generic technology over that technology's life cycle.

There has been limited empirical research to quantify the impact of standards, and other infrastructure technologies, on firms. Link and Tassey (1993) have shown that firms that invest in infrastructure technologies, meaning formulate greater internal capabilities to utilize NIST's output, are more efficient in their in-house R&D than firms that do not. This study, albeit the only one of its kind, follows logically from a theoretical understanding of the role of standards in the innovation process as defined by Tassey (2000, 2005).

12 SMALL BUSINESS INNOVATION RESEARCH PROGRAM

The Small Business Innovation Research Program (SBIR) is a public/private partnership that leverages public R&D through direct governmental support.[1] See Table 12.1.

The Small Business Innovation Research (SBIR) program began at the National Science Foundation (NSF) in 1977 (Tibbetts 1999). At that time the goal of the program was to encourage small businesses, long believed to be engines of innovation in the U.S. economy, to participate in NSF-sponsored research, especially research that had commercial potential. Because of the early success of the program at NSF, Congress passed the Small Business Innovation Development Act of 1982. The Act required all government departments and agencies with external research programs of greater than $100 billion to establish their own SBIR programs and to set aside funds equal to 0.2 percent of the external research budget. As a set aside program, the SBIR program redirects existing R&D rather than appropriating new monies for R&D. Currently, agencies must allocate 2.5 percent of the external research budget to SBIR.

The 1982 Act states that the objectives of the program are:

(1) to stimulate technological innovation
(2) to use small business to meet Federal research and development needs
(3) to foster and encourage participation by minority and disadvantaged persons in technological innovation
(4) to increase private sector commercialization of innovations derived from federal research and development.

---

[1] This chapter draws on Audretsch, Link, and Scott (2002).

The Act was reauthorized in 1992 with the same objectives.

Table 12.1. Taxonomy of Public/Private Partnerships

	Economic Objective	
Governmental Involvement	*Leverage Public R&D*	*Leverage Private R&D*
Indirect		Patent system (Patent Act)
		Tax incentives (R&E tax credit)
		Research joint ventures (NCRA and NCRPA)
Direct		
Financial Resources	**Small Business Innovation Research Program** (Small Business Innovation Development Act)	Advanced Technology Program (Omnibus Trade and Competitiveness Act)
Infrastructural Resources	National Institute of Standards and Technology (Organic Act)	National Institute of Standards and Technology (Organic Act)
Research Resources	National Institute of Standards and Technology (Organic Act)	National Institute of Standards and Technology (Organic Act)

SBIR awards are of three types. Phase I awards are small, generally less than $100,000. The purpose of these awards is to assist firms to assess the feasibility of the research they propose to undertake for the agency in response to the agency's objectives. Phase II awards can range up to $750,000. These awards are for the firm to undertake and complete its proposed research, hopefully leading to a commercializable product or process.

The Department of Defense's (DoD's) SBIR program has been evaluated in some detail (Wessner 2000). It can be concluded that DoD's SBIR program is encouraging commercialization from research that would not have been undertaken without SBIR support. And moreover, the structure of DoD's SBIR program is overcoming reasons for market failure that previously have caused small firms to underinvest in R&D. In fact, Audretsch, Link, and Scott (2002) found that SBIR R&D does lead to commercialization, and that the net social benefits associated with the program's sponsored research are substantial.

As an illustration of SBIR-sponsored programs, consider the following. In 1996, the Department of Defense (DoD) funded, through its SBIR program, a Georgia-based company to conduct Phase II research on hexavalent chrome. Hexavalent chromium is widely used on battleships in the Navy as well as in many industrial applications. It is, however, a known carcinogen and thus creates a toxic waste problem. The U.S. Environmental Protection Agency (EPA) had been aware of this problem but had not yet mandated that it cease being used because no replacement was available. Congress gave DoD an internal directive to find a replacement material. The Georgia-based company won the competition and received an award to develop such a material. The replacement material is based on a thin-film oxide that can be applied to metal. The thin film is sprayed on metal with a flame, and the residual gas contains a replacement molecular coating that performs like hexavalent chromium but is more environmentally friendly.

13 PROGRAM EVALUATION

$\mathcal{F}$undamental to public support of innovation (Chapter 3) is the public sector's awareness of its accountability of its use of public resources. The concept of public accountability can be traced as far back as President Woodrow Wilson's reforms, and in particular to the Budget and Accounting Act of 1921.[1] This Act of June 10, 1921 not only required the President to transmit to Congress a detailed budget on the first day of each regular session, but also it established the General Accounting Office (GAO) to settle and adjust all accounts of the government. We note this fiscal accountability origin because the GAO has had a significant role in the evolution of accountability-related legislation during the past decade.

What follows is a review of the legislative history of initiatives that falls broadly under the rubric of public accountability. As Collins (1997, p. 7) notes:

> As public attention has increasingly focused on improving the performance and accountability of Federal programs, bipartisan efforts in Congress and the White House have produced new legislative mandates for management reform. These laws and the associated Administration and Congressional policies call for a multifaceted approach— including the provision of better financial and performance information for managers, Congress, and the public and the adoption of integrated processes for planning, management, and assessment of results.

---

[1] This section draws from Link and Scott (1998b), Link (1999b), and Link and Scott (2005a).

Fundamental to any evaluation of resources allocation to any program is the recognition that the institution allocating the resources or administering the program is accountable to the public—that is to taxpayers—for its activities. With regards to technology-based institutions, this accountability refers to being able to document and evaluate research performance using metrics that are meaningful to the institutions' stakeholders, meaning to the public.

PERFORMANCE ACCOUNTABILITY

Chief Financial Officers Act of 1990

The GAO has a long-standing interest and a well documented history of efforts to improve governmental agency management through performance measurement. For example, in February 1985 the GAO issued a report entitled "Managing the Cost of Government—Building An Effective Financial Management Structure" which emphasized the importance of systematically measuring performance as a key area to ensure a well-developed financial management structure.

On November 15, 1990, the 101st Congress passed the Chief Financial Officers Act of 1990. As stated in the legislation as background for this Act:

> The Federal Government is in great need of fundamental reform in financial management requirements and practices as financial management systems are obsolete and inefficient, and do not provide complete, consistent, reliable, and timely information.

The stated purposes of the Act are to:

(1) Bring more effective general and financial management practices to the Federal Government through statutory provisions which would establish in the Office of Management and Budget a Deputy Director for Management, establish an Office of Federal Financial Management headed by a Controller, and designate a Chief Financial Officer in each executive department and in each major executive agency in the Federal Government.

    (2) Provide for improvement, in each agency of the Federal Government, of systems of accounting, financial management, and internal controls to assure the issuance of reliable financial information and to deter fraud, waste, and abuse of Government resources.

    (3) Provide for the production of complete, reliable, timely, and consistent financial information for use by the executive branch of the Government and the Congress in the financing, management, and evaluation of Federal programs.

The key phrase in these stated purposes is in point (3) above, "evaluation of Federal programs." Toward this end, the Act calls for the establishment of agency Chief Financial Officers, where agency is defined to include each of the Federal Departments. And, the agency Chief Financial Officer shall, among other things, "develop and maintain an integrated agency accounting and financial management system, including financial reporting and internal controls," which, among other things, "provides for the systematic measurement of performance."

While the Act does outline the many fiscal responsibilities of agency Chief Financial Officers, and their associated auditing process, the Act's only clarification of "evaluation of Federal programs" is in the above phrase, "systematic measurement of performance." However, neither a definition of "performance" nor guidance on "systematic measurement" is provided in the Act. Still, these are the seeds for the growth of attention to performance accountability.

Government Performance and Results Act of 1993

Legislative history is clear that the Government Performance and Results Act (GPRA) of 1993 builds upon the February 1985 GAO report and the Chief Financial Officers Act of 1990. The 103rd Congress stated in the August 3, 1993 legislation that it finds, based on over a year of committee study, that:

    (1) waste and inefficiency in Federal programs undermine the confidence of the American people in the Government and reduce the Federal Government's ability to address adequately vital public needs;

    (2) Federal managers are seriously disadvantaged in their efforts to improve program efficiency and effectiveness,

because of insufficient articulation of program goals and inadequate information on program performance; and

(3) congressional policymaking, spending decisions and program oversight are seriously handicapped by insufficient attention to program performance and results.

Accordingly, the purposes of GPRA are to:

(1) improve the confidence of the American people in the capability of the Federal Government, by systematically holding Federal agencies accountable for achieving program results;

(2) initiate program performance reform with a series of pilot projects in setting program goals, measuring program performance against those goals, and reporting publicly on their progress;

(3) improve Federal program effectiveness and public accountability by promoting a new focus on results, service quality, and customer satisfaction;

(4) help Federal managers improve service delivery, by requiring that they plan for meeting program objectives and by providing them with information about program results and service quality;

(5) improve congressional decisionmaking by providing more objective information on achieving statutory objectives, and on the relative effectiveness and efficiency of Federal programs and spending; and

(6) improve internal management of the Federal Government.

The Act requires that the head of each agency submit to the Director of the Office of Management and Budget (OMB):

> ... no later than September 30, 1997 ... a strategic plan for program activities. Such plan shall contain ... a description of the program evaluations used in establishing or revising general goals and objectives, with a schedule for future program evaluations.

And, quite appropriately, the Act defines program evaluation to mean:

> ... an assessment, through objective measurement and systematic analysis, of the manner and extent to which federal programs achieve intended objectives.

In addition, each agency is required to:

> ... prepare an annual performance plan [beginning with fiscal year 1999] covering each program activity set forth in the budget of such agency. Such plan shall ... establish performance indicators to be used in measuring or assessing the relevant outputs, service levels, and outcomes of each program activity;

where "performance indicator means a particular value or characteristic used to measure output or outcome."

Cozzens (1995) speculated, at the time of the passage of GPRA, that it will encourage agencies to ignore what is difficult to measure, no matter how relevant. Alternatively, one could wear a more pessimistic hat and state that GPRA will encourage agencies to emphasize what is easy to measure, no matter how irrelevant.

FISCAL ACCOUNTABILITY

Legislation following GPRA emphasizes fiscal accountability more than performance accountability. While it is not our intent to suggest that performance accountability is more or less important than fiscal accountability, for we believe that both aspects of public accountability are important, our emphasis below is on performance accountability. Nevertheless, our discussion would not be complete in this chapter without references to the Government Management Reform Act of 1994 and the Federal Financial Management Improvement Act of 1996.

Government Management Reform Act of 1994

The Government Management Reform Act of 1994 builds on the Chief Financial Officers Act of 1990. Its purpose is to improve the management of the federal government though reforms to the management of federal human resources and financial management. Motivating the Act is the belief that federal agencies must streamline their operations and must rationalize their resources to better match a growing demand on their services.   Government, like the private sector, must adopt modern management methods, utilize meaningful program performance measures, increase workforce incentives without sacrificing accountability, and strengthen the overall delivery of services.

Federal Financial Management Improvement Act of 1996

The Federal Financial Management Improvement Act of 1996 follows from the belief that federal accounting standards have not been implemented uniformly through federal agencies. Accordingly, this Act establishes a uniform accounting reporting system in the federal government.

The above overview of what we call public accountability legislation makes clear that government agencies are becoming more and more accountable for their fiscal and performance actions. And, these agencies are being required to a greater degree than ever before to account for their activities through a process of systematic measurement. For technology-based institutions in particular, internal difficulties are arising as organizations learn about this process.

Compliance with these guidelines is causing increased planning and impact assessment activity and is also stimulating greater attention to methodology. Perhaps there is no greater validation of this observation than the diversity of response being seen among public agencies, in general, and technology-based public institutions, in particular, as they grope toward an understanding of the process of documenting and assessing their public accountability.   Activities in recent years have ranged from interagency discussion meetings to a reinvention of the assessment wheel, so to speak, in the National Science and Technology Council's (1996) report, "Assessing Fundamental Science."

SYSTEMATIC APPROACHES TO THE EVALUATION OF PUBLIC/PRIVATE PARTNERSHIPS

GPRA is directionally, as opposed to methodologically, clear about the evaluation process. It stipulates that public institutions/research programs must identify outputs and quantify the economic benefits of the outcomes associated with such outputs. In our opinion, agencies will attempt to quantify outcome benefits and then compare those quantified benefits to the public costs to achieve the benefits. Although these are GPRA's directions, the methodological hurdle that has been plaguing most public agencies is how to quantify benefits. And even with an acceptable quantification of benefits, will the confidence of the American people in public sector research be strengthened by simply comparing benefits to costs?

Consider two different approaches to program evaluation. When evaluating publicly-funded publicly-performed research, the relevant approach is based on a counterfactual method; when evaluating publicly-funded privately-performed research programs, the relevant approach is based on a spillover method. The discussion that follows draws upon Link and Scott (1998b, 2000, forthcoming a).

Traditional Evaluation Methods

Griliches (1958) and Mansfield, et al. (1977) pioneered the application of fundamental economic insight to the development of estimates of private and social rates of return to investments in R&D. Streams of investment outlays through time—the costs—generate streams of economic surplus through time—the benefits. Once identified and measured, these streams of costs and benefits are used to calculate rates of return, benefit-to-cost ratios, or other related metrics.

In the Griliches/Mansfield model, the innovations evaluated are conceptualized as reducing the cost of producing a good sold in a competitive market at constant unit cost. For any period, there is a demand curve for the good and a horizontal supply curve. Innovation lowers the unit cost of production, shifting downward the horizontal supply curve and thereby, at the new lower equilibrium price, resulting in greater consumer surplus (economists' measure of value—the difference between the price consumers would have been willing to pay and the actual price, summed over all purchases, they paid). Additionally, the Griliches/Mansfield model accounts for producer surplus, measured as the difference between the price the producers receive and the actual marginal cost, summed over the output sold, minus any fixed costs. Social benefits are then the streams of new consumer and producer surpluses, while private benefits are the streams of producer surplus, not all of which are necessarily new because the surplus gained by one producer may be cannibalized from the pre-innovation surplus of another producer. Social and private costs will, in general, also be divergent.

The Griliches/Mansfield model for calculating economic social rates of return add the public and the private investments through time to determine social investment costs, and then the stream of new economic surplus generated from those investments is the benefit. Thus, the evaluation question that can be answered from such an evaluation analysis is: What is the social rate of return to the innovation, and how does that compare to the private rate of return? We argue that this is not the most appropriate question to ask from a public accountability perspective. The fact that the social rate of return is greater than the private rate of return may validate the role of government in innovation if the private sector would not have undertaken the research; but it ignores, for example, consideration of the cost effectiveness of the public sector undertaking the research as opposed to the private sector.

The Counterfactual Evaluation Method

A different question should be considered when publicly-funded publicly-performed investments are evaluated. Holding constant the very stream of economic surplus that the Griliches/Mansfield model seeks to measure, and making no attempt to measure that stream, one should ask the counterfactual question: What would the private sector have had to invest to achieve those benefits in the absence of the public sector's investments? The answer to this question gives the benefits of the public's investments—namely, the costs avoided by the private sector. With those benefits—obtained in practice through extensive interviews with administrators, federal research scientists, and those in the private sector who would have to duplicate the research in the absence of public performance—counterfactual rates of return and benefit-to-cost ratios can be calculated to answer the fundamental evaluation question: Are the public investments a more efficient way of generating the technology than private sector investments would have been? The answer to this question is more in line with the public accountability issues implicit in GPRA, and certainly is more in line with the thinking of public sector stakeholders, who may doubt the appropriateness of government's having a role in the innovation process in the first place.

The Spillover Evaluation Method

There are important projects where economic performance can be improved with public funding of privately-performed research. Public funding is needed when socially valuable projects would not be undertaken without it. If the expected private rate of return from a research project falls short of the required rate called the hurdle rate, then the private sector firm will not invest in the project. Nonetheless, if the benefits of the research spill over to consumers and to firms other than the ones investing in the research, the social rate of return may exceed the appropriate hurdle rate. It would then be socially valuable to have the investments made, but since the private investor will not make them, the public sector should. By providing some public funding, thereby reducing the investment amount needed from the private firm or firms doing the research, the expected private rate of return can be increased above the hurdle rate. Thus, because of this subsidy, the private firm is willing to perform the research which is socially desirable because much of its output spills over to other firms and sectors in the economy.

The question asked in the spillover method is one that facilitates an economic understanding of whether the public sector should be underwriting a portion of private-sector firms' research, namely: What proportion of the total profit stream *generated by the private firm's R&D and innovation* does the private firm expect to capture; and hence, what proportion is not appropriated but is instead captured by other firms that imitate the innovation or use knowledge generated by the R&D to produce competing products for the social good? The part of the stream of expected profits captured by the innovator is its private return, while the entire stream is the lower bound on the social rate of return. In essence, this method weighs the private return, estimated through extensive interviews with firms receiving public support regarding their expectations of future patterns of events and future abilities to appropriate R&D-based knowledge, against private investments. The social rate of return weights the social returns against the social investments.

The application of the spillover model to the evaluation of public funding/private performance of research is appropriate since the output of the research is only partially appropriable by the private firm with the rest spilling over to society. The extent of the spillover of such knowledge with public good characteristics determines whether or not the public sector should fund or partially fund the research.

EVALUATION METRICS

For an evaluation of a public/private partnership, two time series of data are needed. One time series is on the costs allocated to the partnership, public costs and private costs. The other time series is on the benefits to those whose R&D is being leveraged, measured in constant dollars. Several metrics are common for estimating the social value of the program.

Internal Rate of Return

The internal rate of return (IRR) is the value of the discount rate, i, that equates the net present value (NPV) of the stream of net benefits associated with a research project to zero. The time series runs from the beginning of the research project, $t = 0$, to a terminal point, $t = n$.

Mathematically,

(1) $NPV = [(B_0 - C_0) / (1 + i)^0] + \ldots + [(B_n - C_n) / (1 + i)^n] = 0$

where, $(B_t - C_t)$ represents the net benefits associated with the project in year t, and n represents the number of time periods – years in the case study evaluated in this paper – being considered in the evaluation.

For unique solutions for i, from equation (1), the IRR can be compared to a value, r, that represents the opportunity cost of funds invested by the technology-based public institution. Thus, if the opportunity cost of funds is less than the internal rate of return, the project was worthwhile from an *ex post* social perspective.

Benefit-to-Cost Ratio

The ratio of benefits-to-costs (B / C) is the ratio of the present value of all measured benefits to the present value of all measured costs. Both benefits and costs are referenced to the initial time period, $t = 0$, when the project began as:

(2) $B / C = [\sum_{t=0 \text{ to } t=n} B_t / (1 + r)^t] / [\sum_{t=0 \text{ to } t=n} C_t / (1 + r)^t]$

A benefit-to-cost ratio of 1 is said to indicate a project that breaks-even. Any project with B / C > 1 is a relatively successful project as defined in terms of benefits exceeding costs.

Fundamental to implementing the ratio of benefits-to-costs is a value for the discount rate, r. While the discount rate representing the opportunity cost for public funds could differ across a portfolio of public investments, the calculated metrics in this paper follow the guidelines set forth by the Office of Management and Budget (1992), which states that: "Constant-dollar benefit-cost analyses of proposed investments and regulations should report net present value and other outcomes determined using a real discount rate of 7 percent."

Net Present Value

The information developed to determine the benefit-to-cost ratio can be used to determine net present value (NPV) as:

(3) $NPV_{initial\ year} = B - C$

where, as in the calculation of B / C, B refers to the present value of all measured benefits and C refers to the present value of all measured costs and where present value refers to the initial year or time period in which the project began, $t = 0$ in terms of the B / C formula in equation (2). Note that NPV allows, in principle, one means of ranking several projects *ex post*, providing investment sizes are similar.

To compare the net present values across different case studies with different starting dates, the net present value for each can be brought forward to the same year – here year 2002. The $NPV_{initial\ year}$ is brought forward under the assumption that the NPV for the project was invested at the 7 percent real rate of return that is recommended by the Office of Management and Budget as the opportunity cost of government funds. NPV_{2002} is then a project's NPV multiplied by 1.07 raised to the power of 2002 minus the year that the project was initiated as:

(4) $NPV_{2002} = NPV\ x\ (1.07)^{2002-initial\ year}$

14 CONCLUDING STATEMENT

This book has focused on six specific public/private partnerships in the United States, what they are and their role in stimulating innovative activity. These six partnerships are summarized in Table 14.1. Certainly, public/private partnerships are not unique to the United States, although the pedagogical focus of the book is general enough to be applied to partnerships in any industrialized nation. Regardless, certain fundamental principles remain, and they are pervasive throughout the book:

- Government has a role in the innovation process and that role is based on the concept of market failure.
- Ideally, the causes of market failure should be understood *a priori*, and then an appropriate innovation policy proffered and adopted—a public/private partnership—to alleviate the barriers that caused the market failure.[1]
- Whatever innovation policy is implemented, it is incumbent on the government to demonstrate, at some point in time, accountability for its use of public resources. And, as the summary tables throughout the book have emphasized, any effort at accountability must take into account the policy's ability to leverage public and private R&D.

The models of economic growth from Chapter 5 are reproduced in this chapter as Figure 14.1 and Figure 14.2 to serve as summary devices. The six public/private partnerships discussed herein map into the models of economic growth.

---

[1] Note the word "ideally" in this bulleted conclusion since in practice innovation and technology policies are generally justified *ex post*. They are then rationalized after the fact on the assumption that there was a market failure that needed to be overcome. The notable exception is in Link and Scott (1998a).

Table 14.1. Taxonomy of Public/Private Partnerships

	Economic Objective	
Governmental Involvement	*Leverage Public R&D*	*Leverage Private R&D*
Indirect		**Patent system** (Patent Act)
		Tax incentives (R&E tax credit)
		Research joint ventures (NCRA and NCRPA)
Direct		
Financial Resources	**Small Business Innovation Research Program** (Small Business Innovation Development Act)	**Advanced Technology Program** (Omnibus Trade and Competitiveness Act)
Infrastructural Resources	**National Institute of Standards and Technology** (Organic Act)	**National Institute of Standards and Technology** (Organic Act)
Research Resources	**National Institute of Standards and Technology** (Organic Act)	**National Institute of Standards and Technology** (Organic Act)

For specific examples, although the mapping is not singularly focused, the patent system leverages the relationship between proprietary technologies and technology development in manufacturing, while it constrains the relationship between purchased technologies and entrepreneurial activity in services when the purchased technology was patented by the inventive firm.

Tax incentives related to R&D affect the level of investment in generic technologies and proprietary technologies in manufacturing, and they affect in services the so-called make versus by decision (in-house technology development versus purchased technologies).

Figure 14.1. Entrepreneurial Model of Innovation in a Technology-based Manufacturing Sector Firm

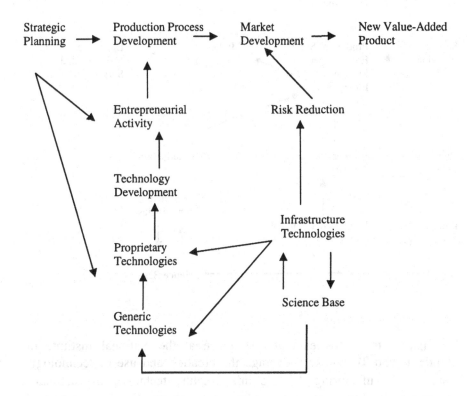

In manufacturing, research joint ventures affect the speed at which activities occur within the development of generic technologies and the level of spillover knowledge that carries over into the development of proprietary technologies. In the service sector, there is relatively little joint venture activity.

Firms in all sectors can and do benefit from the Small Business Innovation Research Program, although the lion's share of Department of Defense (DoD, highlighted in Chapter 12) awards have gone to manufacturing firms. The same is true for awards from the Advanced

Technology Program. In both cases, with reference to Figure 14.1, generic technology development is enhanced and the relationship between generic technology and proprietary technology is leveraged.

Figure 14.2. Entrepreneurial Model of Innovation in a Technology-based Service Sector Firm

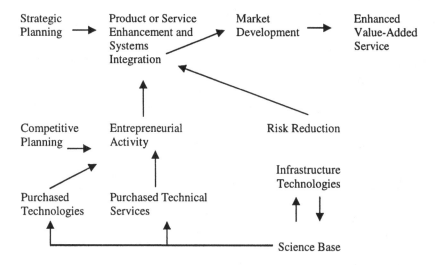

Finally, the activities that take place at the National Institute of Standards and Technology leverage the creation and use of technology. Thus, in manufacturing, NIST's infrastructure technologies—standards and protocols—enrich the science base and enhance the development of generic technology and proprietary technology, thus affecting risk reduction. In services, infrastructure technologies primarily affect the science base and risk reduction.

Albeit a simplistic summarization, it does underscore the breadth of influence that public/private partnership have on the innovation process. Reemphasizing the conceptual framework from Chapter 5:

R&D → Knowledge → Innovation → Technological Advancement →
Economic Growth

public/private partnerships affect R&D activity and thus innovation. Innovation in turn leads to technological advancement, and technological advancement leads to economic growth. Thus, public/private partnerships encompass many policy alternatives that are part of a Nation's innovation strategy.

REFERENCES

Abramovitz, Moses. "Resource and Output Trends in the United States since 1870," *American Economic Review*, 1956, pp. 5-23.

Acs, Zoltan, and David B. Audretsch. "Patents as a Measure of Innovative Activity," *Kyklos*, 1989, pp. 171-180.

Acs, Zoltan and David B. Audretsch. *Innovation and Small Firms*, Cambridge: MIT Press, 1990.

Acs, Zoltan, David B. Audretsch, and Maryann P. Feldman. "Real Effects of Academic Research: Comment," *American Economic Review*, 1992, pp. 363-367.

Acs, Zoltan, David B. Audretsch, and Maryann P. Feldman. "R&D Spillovers and Recipient Firm Size," *Review of Economics and Statistics*, 1994, pp. 336-340.

Adams, James D. "Comparative Localization of Academic and Industrial Spillovers," *Journal of Economic Geography*, 2002, pp. 253-278.

Adams, James D. and Adam B. Jaffe. "Bounding the Effects of R&D: An Investigation Using Matched Establishment-Firm Data," *Rand Journal of Economics*, 1996, pp. 700-721.

Almeida, Paul and Bruce Kogut. "The Exploitation of Technological Diversity and the Geographical Localization of Innovation," *Small Business Economics*, 1997, pp. 21-31.

Arrow, Kenneth. J. "Economic Welfare and the Allocation of Resources for Invention," in *The Rate and Direction of Inventive Activity*, Princeton: Princeton University Press, 1962, pp. 609-625.

Arrow, Kenneth. J. *Social Choice and Individual Values*, New Haven: Yale University Press, 1963.

Association of University Related Research Parks (AURRP). "Worldwide Research & Science Park Directory 1998," BPI Communications, 1997.

Audretsch, David B. *Innovation and Industry Evolution*, Cambridge: MIT Press, 1995.

Audretsch, David B., Barry Bozeman, Kathryn L. Combs, Maryann P. Feldman, Albert N. Link, Donald S. Siegel, Paula E. Stephan, Gregory Tassey, and Charles Wessner. "The Economics of Science and Technology," *Journal of Technology Transfer*, 2002a, pp. 155-203.

Audretsch, David B. and Maryann P. Feldman. "R&D Spillovers and the Geography of Innovation and Production," *American Economic Review*, 1996, pp. 630-640.

Audretsch, David B., Albert N. Link, and John T. Scott. "Public/Private Partnerships: Evaluating SBIR-Supported Research," *Research Policy*, 2002b, pp. 145-158.

Audretsch, David B. and Paula E. Stephan. "Company-Scientist Locational Links: The Case of Biotechnology," *American Economic Review*, 1996, pp. 641-652.

Austin, David H. "An Event Study Approach to Measuring Innovative Output: The Case of Biotechnology," *American Economic Review Papers and Proceedings*, 1993, pp. 253-258.

Baldwin, William L., Scott, John T. *Market Structure and Technological Change*, London: Harwood Academic Publishers, 1987.

Barnett, Homer G. *Innovation: The Basis of Cultural Change*, New York, W.W. Norton, 1953.

Baudeau, Nicolas. *Premier Introduction à la Philosophie Économique*, edited by A. Dubois, Paris: P, Geuthner, 1910 [originally 1767].

Ben-Zion, Uri. "The R&D and Investment Decision and its Relationship to the Firm's Market Value: Some Preliminary Results," in *R&D, Patents, and Productivity* (edited by Z. Griliches), Chicago: University of Chicago Press, 1984, pp. 134-162.

Bound, John, Clint Cummins, Zvi Griliches, Bronwyn Hall, and Adam Jaffe. "Who Does R&D and Who Patents?," in *R&D, Patents, and Productivity* (edited by Z. Griliches), Chicago: University of Chicago Press, 1984, pp. 21-54.

Bozeman, Barry. "Technology Transfer and Public Policy: A Review of Research and Theory," *Research Policy*, 2000: pp. 627-656.

Bozeman, Barry. and Michael Crow. "Red Tape and Technology Transfer in U.S. Government Laboratories," *Journal of Technology Transfer*, 1991, pp. 29-37.

Bozeman, Barry and Albert N. Link. *Investments in Technology: Corporate Strategies and Public Policy Alternatives*, New York: Praeger Publishers, 1983.

Bozeman, Barry and Albert N. Link, "Tax Incentives for R&D: A Critical Evaluation," *Research Policy*, 1984, pp. 21-31.

Brod, Andrew C. and Albert N. Link. "Trends in Cooperative Research Activity: Has the National Cooperative Research Act Been Successful?" in *Innovation Policy in the Knowledge-Based Economy* (edited by M. Feldman and A. Link), Norwell, MA: Kluwer Academic Publishers, 2001, pp. 105-119.

Bush, Vannevar. *Science—the Endless Frontier*. Washington, DC: U.S. Government Printing Office, 1945.

Carr, Robert K. "U.S. Federal Laboratories and Technology Transfer," mimeograph, 1995.

Clark, Burton R. *Places of Inquiry*, Berkeley: University of California Press, 1995.

Coburn, Christopher. Partnerships: A Compendium of State and Federal Technology Programs, Columbus, OH: Battelle Press, 1995.

Cockburn, Iain and Zvi Griliches. "Industry Effects and Appropriability Measures in the Stock Market's Valuation of R&D and Patents," *American Economic Review Papers and Proceedings*, 1988, pp. 419-423.

Cohen, Wesley M. "Thoughts and Questions on Science Parks," presented at the National Science Foundation Science Parks Indicators Workshop, University of North Carolina at Greensboro, November 2002. (Also in A.N. Link, Final Report to the National Science Foundation on Science Park Indicators Workshop, January 2003.)

Cohen, Wesley M. and David A. Levinthal. "Innovation and Learning: The Two Faces of R&D," *Economic Journal*, 1989, pp. 569-596.

Collins, Eileen. "Performance Reporting in Federal Management Reform," National Science Foundation Special Report, mimeographed, 1997.

Council on Competitiveness. *Endless Frontiers, Limited Resources: U.S. R&D Policy for Competitiveness*, Washington, DC: Council on Competitiveness, 1996.

Cozzens, Susan E. "Assessment of Fundamental Science Programs in the Context of the Government Performance and Results Act (GPRA)," Washington, DC: Critical Technologies Institute, 1995.

Crow, Michael and Barry Bozeman. *Limited by Design: R&D Laboratories in the U.S. National Innovation System*, New York: Columbia University Press, 1998.

Danilov, V.J. "The Research Park Shake-Out, *Industrial Research*, 1971, pp. 1-4.

David, Paul A. "Some New Standards for the Economics of Standardization in the Information Age," in *Economic Policy and Technological Performance* (edited by P. Dasgupta and P. Stoneman), Cambridge: Cambridge University Press, 1987, pp. 206-239.

David, Paul A., Bronwyn H. Hall, and Andrew A. Toole. "Is Public R&D a Complement or Substitute for Private R&D? A Review of the Econometric Evidence," *Research Policy*, 2000, pp. 497-529.

Dunne, Timothy. "Plant Age and Technology Usage in U.S. Manufacturing Industries," *Rand Journal of Economics*, 1994, pp. 488-499.

Economic Report of the President. Washington, DC: Government Printing Office, 1994.

Economic Report of the President. Washington, DC: Government Printing Office, 1995.

Eden, Lorraine, Edward Levitas, and Richard J. Martinez. "The Production, Transfer and Spillover of Technology: Comparing Large and Small Multinationals as Technology Producers," *Small Business Economics*, 1997, pp. 53-66.

Executive Office of the President. U.S. Technology Policy, Washington, DC: Office of Science and Technology Policy, 1990.

Feldman, Maryann P. "Knowledge Complementarity and Innovation," *Small Business Economics*, 1994, pp. 363-372.

Feldman, Maryann P. "The New Economics of Innovation, Spillovers and Agglomeration: A Review of Empirical Studies," *Economics of Innovation and New Technology*, 1999, pp. 5-25.

Feldman, Maryann P., Albert N. Link, and Donald S. Siegel. *The Economics of Science and Technology*, Boston: Kluwer Academic Publishers, 2000.

Gallaher, Michael P., Albert N. Link, and Jeffrey E. Petrusa. *Innovation in the U.S. Service Sector*, London: Routledge, forthcoming.

Glaeser, Edward, Heidi Kallal, Jose Scheinkman, and Andrei Shleifer. "Growth of Cities," *Journal of Political Economy*, 1992, pp. 1126-1152.

Grabowski, Henry G. "The Determinants of Industrial Research and Development: A Study of the Chemical, Drug, and Petroleum Industries," *Journal of Political Economy*, 1968, pp. 156-159.

Griliches, Zvi. "Research Costs and Social Returns: Hybrid Corn and Related Innovations," *Journal of Political Economy*, 1958, pp. 501-522.

Griliches, Zvi. "Issues in Assessing the Contribution of R&D to Productivity Growth," *Bell Journal of Economics*, 1979, pp. 92-116.

Griliches, Zvi. "Market Value, R&D, and Patents," *Economics Letters*, 1981, pp. 183-187.

Griliches, Zvi. "Productivity, R&D, and Basic Research at the Firm Level in the 1970s," *American Economic Review*, 1986, pp. 141-154.

Griliches, Zvi. "Patent Statistics as Economic Indicators: A Survey," *Journal of Economic Literature*, 1990, pp. 1661-1707.

Hagedoorn, John, Albert N. Link, and Nicholas S. Vonortas. "Research Partnerships," *Research Policy*, 2000, pp. 567-586.

Hall, Bronwyn H. "Exploring the Patent Explosion," *Journal of Technology Transfer*, 2005, pp. 35-48.

Hall, Bronwyn H., Zvi Griliches, and Jerry A. Hausman. "Patents and R&D: Is There a Lag?" *International Economic Review*, 1986, pp. 265-302.

Hall, Bronwyn H., Adam Jaffe and Manuel Trajtenberg. "Market Value and Patent Citations: A First Look" National Bureau of Economic Research Working Paper No. 7741, June 2000.

Hall, Bronwyn H., Albert N. Link, and John T. Scott. "Universities as Research Partners," NBER Working Paper No. 7643, April 2000.

Hall, Bronwyn H., Albert N. Link, and John T. Scott, "Barriers Inhibiting Industry from Partnering with Universities: Evidence from the Advanced Technology Program," *Journal of Technology Transfer*, 2001, pp. 87-98.

Hall, Bronwyn H., Link, Albert N., Scott, John T. "Universities as Research Partners," *Review of Economics and Statistics*, 2003, pp. 485-491.

Hall, Bronwyn H. and John van Reenen. "How Effective are Fiscal Incentives for R&D? A Review of the Evidence," *Research Policy*, 2000, pp. 449-469.

Harrod, Roy F. *Toward a Dynamic Economics*, London: Macmillan, 1948.

Hébert, Robert F. and Albert N. Link. *The Entrepreneur: Mainstream Views & Radical Critiques*, second edition, New York: Praeger Publishers, 1988.

Hébert, Robert F. and Albert N. Link. "In Search of the Meaning of Entrepreneurship," *Small Business Economics*, 1989, pp. 39-49.

Henderson, Rebecca, Adam Jaffe and Manuel Trajtenberg, "Universities as a Source of Commercial Technology: A Detailed Analysis of University Patenting, 1965-1988," *Review of Economics and Statistics*, 1998, pp. 119-127.

Hicks, John R., *Theory of Wages*, London: Macmillan, 1932.

Hounshell, David A. "The Evolution of Industrial Research in the United States," in *Engines of Innovation: U.S. Industrial Research at the End of an Era* (edited by R. Rosenbloom and W. Spencer), Boston: Harvard Business School Press, 1996, pp. 13-86.

House Committee on Science. *Unlocking Our Future: Toward a New National Science Policy*, Report to Congress, September 24, 1998.

Jaffe, Adam B. "Technological Opportunity and Spillovers of R&D: Evidence from Firm's Patents, Profits, and Market Value," *American Economic Review*, 1986, pp. 984-1001.

Jaffe, Adam B. "Real Effects of Academic Research," *American Economic Review*, 1989, pp. 957-970.

Jaffe, Adam B. "The Importance of 'Spillovers' in the Policy Mission of the ATP," *Journal of Technology Transfer*, 1998, pp. 11-19.

Jaffe, Adam B., Michael S. Fogarty, and Bruce A. Banks. "Evidence From Patents and Patent Citations on the Impact of NASA and other Federal Labs on Commercial Innovation," *Journal of Industrial Economics*, 1998, pp. 183-206.

Jaffe, Adam, Manuel Trajtenberg, and Rebecca Henderson. "Geographical Localization of Knowledge Spillovers as Evidenced by Patent Citations," *Quarterly Journal of Economics*, 1993, pp. 577-598.

Kaufer, Erich. *The Economics of the Patent System*, Chur, UK, Harwood Academic Publishers, 1989.

Kerr, Clark. *Troubled Times in American Higher Education*, Albany: State University of New York Press, 1994.

Kortum, Samuel and Josh Lerner. "What is Behind the Recent Surge in Patenting?" *Research Policy*, 1999, pp. 1-22.

Kuznets, Simon. "Inventive Activity: Problems of Definition and Measurement," in The Rate and Direction of Inventive Activity (edited by R. Nelson), Princeton: National Bureau of Economic Research, 1962.

Leyden, Dennis P. and Albert N. Link. "Tax Policies Affecting R&D: An International Comparison," *Technovation*, 1993, pp. 17-25.

Leyden, Dennis P. and Albert N. Link. "Federal Laboratories as Research Partnerships," *International Journal of Industrial Organization*, 1999, pp. 575-592.

Lichtenberg, Frank R. and Donald Siegel. "The Impact of R&D Investment of Productivity-New Evidence Using Linked R&D-LRD Data," *Economic Inquiry*, 1991, pp. 203-228.

Link, Albert N. "Basic Research and Productivity Increase in Manufacturing," *American Economic Review*, 1981, pp. 1111-1112.

Link, Albert N. "Market Structure and Voluntary Product Standards," *Applied Economics*, 1983, pp. 393-401.

Link, Albert N. *Technological Change and Productivity Growth*, London: Harwood Academic Publishers, 1987.

Link, Albert N. *A Generosity of Spirit: The Early History of Research Triangle Park*, Research Triangle Park, NC: Research Triangle Foundation of North Carolina, 1995.

Link, Albert N. "Economic Impact Assessment Guidelines for Conducting and Interpreting Assessment Studies," NIST Planning Report 96-1, 1996a.

Link, Albert N. *Evaluating Public Sector Research and Development*, New York, Praeger Publishers, 1996b.

Link, Albert N. "On the Classification of Industrial R&D," *Research Policy*, 1996c, pp. 397-401.

Link, Albert N. "Public/Private Partnerships as a Tool to Support Industrial R&D: Experiences in the United States," final report prepared for the working group on technology and innovation policy, OECD Division of Science and Technology, 1998.

Link, Albert N. "Public/Private Partnerships in the United States," *Industry and Innovation*, 1999a, pp. 191-217.

Link, Albert N. "A Primer on the Economics of Science and Technology," University of North Carolina at Greensboro working paper, 1999b.

Link, Albert N. *From Seed to Harvest: The Growth of Research Triangle Park*, Research Triangle Park, NC: Research Triangle Foundation of North Carolina, 2002.

Link, Albert N. and Laura L. Bauer. *Cooperative Research in U.S. Manufacturing*, Lexington, MA: Lexington Books, 1989.

Link, Albert N. and Jamie R. Link. "Women in Science: An Exploratory Analysis of Trends in the United States," *Science and Public Policy*, 1999, pp. 437-442.

Link, Albert N. and Kevin R. Link. "On the Growth of U.S. Science Parks," *Journal of Technology Transfer*, 2003, 81-85, 2003.

Link, Albert N. and John Rees. "Firm Size, University Based Research and the Returns to R&D," *Small Business Economics*, 1990, pp. 25-32.

Link, Albert N. and John T. Scott. "Assessing the Infrastructural Needs of a Technology-Based Service Sector: A New Approach to Technology Policy Planning," *STI Review*, 1998a, pp. 171-204.

Link, Albert N. and John T. Scott. *Public Accountability: Evaluating Technology-Based Institutions*, Norwell, MA: Kluwer Academic Publishers, 1998b.

Link, Albert N. and John T. Scott. "Evaluating Federal Research Programs," mimeographed, 2000.

Link, Albert N. and John T. Scott. "Public/Private Partnerships: Stimulating Competition in a Dynamic Economy," *International Journal of Industrial Organization*, 2001, pp. 763-794.

Link, Albert N. and John T. Scott. "Explaining Observed Licensing Agreements: Toward a Broader Understanding of Technology Flows," *Economics of Innovation and New Technology*, 2002, pp. 211-231.

Link, Albert N. and John T. Scott. "The Growth of Research Triangle Park," *Small Business Economics*, 2003a, pp. 167-175.

Link, A.N. and John T. Scott. "U.S. Science Parks: The Diffusion of an Innovation and Its Effects on the Academic Mission of Universities," *International Journal of Industrial Organization*, 2003b, pp. 1323-1356.

Link, Albert N. and John T. Scott. "Evaluation of ATP's Intramural Research Awards Program," final report to the National Institute of Standards and Technology, Advanced Technology Program, 2004.

Link, Albert N. and John T. Scott. "Evaluating Public Sector R&D Programs: The Advanced Technology Program's Investment in Wavelength References for Optical Fiber Communications," *Journal of Technology Transfer*, 2005a, pp. 241-251.

Link, Albert N. and John T. Scott. "Universities as Research Partners in U.S. Research Joint Ventures," *Research Policy*, 2005b, pp. 385-393.

Link, Albert N. and John T. Scott. *Evaluating Public Research Institutions: The U.S. Advanced Technology Program's Intramural Research Initiative*, London: Routledge, forthcoming a.

Link, Albert N. and John T. Scott. "An Economic Evaluation of the Baldrige National Quality Program," *Economics of Innovation and New Technology*, forthcoming b.

Link, Albert N. and John T. Scott. "U.S. University Research Parks," *Journal of Productivity Analysis*, forthcoming c.

Link, Albert N. and Donald S. Siegel. *Technolgical Change and Economic Performance*, London: Routledge, 2003.

Link, Albert N. and Gregory Tassey. *Strategies for Technology-based Competition: Meeting the New Global Challenge*, Lexington, MA: Lexington Books, 1987.

Link, Albert N. and Gregory Tassey. "The Technology Infrastructure of Firms: Investments in Infratechnology," *IEEE Transactions on Engineering Management*, 1993, pp. 312-315.

Link, Albert N., David J. Teece, and William Finan. "Estimating the Benefits from Collaboration: The Case of SEMATECH, *Review of Industrial Organization*, 1996, pp. 737-751.

Link, Albert N. and Robert Zmud. "External Sources of Technical Knowledge," *Economics Letters*, 1987, pp. 295-299.

Machlup, Fritz. *Knowledge and Knowledge Production*, Princeton: Princeton University Press, 1980.

Mansfield, Edwin. *Industrial Research and Technological Change*, New York: W.W., Norton, 1968.

Mansfield, Edwin. *Technological Change*, New York: W.W. Norton, 1971.

Mansfield, Edwin. "Basic Research and Productivity Increase in Manufacturing," *American Economic Review*, 1980, pp. 863-873.

Mansfield, Edwin, John Rapoport, Anthony Romeo, Samuel Wagner, and George Beardsley. "Social and Private Rates of Return from Industrial Innovations," *Quarterly Journal of Economics*, 1977, pp. 221-240.

Martin, Stephen and John T. Scott. "The nature of Innovation Market Failure and the Design of Public Support for Private Innovation," *Research Policy*, 2000, pp. 437-447.

Mises, Ludwig von. *Human Action: A Treatise on Economics*, New Haven: Yale University Press, 1949.

Morgan, Robert P. "University Research Contributions to Industry: The Faculty View," in *Trends in Industrial Innovation: Industry Perspectives & Policy Implications* (edited by P. Blain and R. Frosch), Research Triangle Park: Sigma Xi, The Scientific Research Society, 1998, pp. 163-170.

Morrison, Catherine and Donald Siegel. "External Capital Factors and Increasing Returns in U.S. Manufacturing," *Review of Economics and Statistics*, 1997, pp. 647-654.

Mowery, David C. and David J. Teece. "Strategic Alliances and Industrial Research," in *Engines of Innovation: U.S. Industrial Research at the*

End of an Era (edited by R. Rosenbloom and W. Spencer), Boston: Harvard Business School Press, 1996, pp. 111-129.

Mueller, Dennis C. "The Firm Decision Process: An Econometric Investigation," *Journal of Political Economy*, 1967, pp. 58-87.

National Research Council. *Standards, Conformity Assessment, and Trade,* Washington, DC: National Academy Press, 1995.

National Resources Committee. *Research—A National Resource.* Washington, DC: Science Committee of the National Resources Committee, November 1938.

National Science and Technology Council. "Assessing Fundamental Science," Washington, DC: National Science and Technology Council, 1996.

National Science Board. *Science & Engineering Indicators*, Washington, DC: National Science Foundation, bi-annual.

National Science Foundation. *Graduate Studies and Postdoctorates in Science and Engineering: 1996*, NSF 98-301. Washington, DC: National Science Foundation, 1998.

Nekar, Atul and Scott Shane. "When Do Start-Ups that Exploit Patented Academic Knowledge Survive," *International Journal of Industrial Organization*, 2003, pp. 1391-1410, 2003.

Nelson, Richard R. *National Innovation Systems: A Comparative Analysis*, Oxford: Oxford University Press, 1993.

Nightingale, Paul. "A Cognitive Model of Innovation," *Research Policy*, 1998, pp. 689-709.

Office of Management and Budget. "Circular No. A-94: Guidelines and Discount Rates for Benefit-Cost Analysis of Federal Programs," Washington, DC, 1992.

Office of Management and Budget. "Economic Analysis of Federal Regulations under Executive Order 12866," Washington, DC, 1996.

Office of the President, *U.S. Technology Policy*, Washington, DC: Executive Office of the President, September, 26, 1990.

Office of Science and Technology Policy. "Science in the National Interest," Washington, DC: Executive Office of the President, 1994.

Office of Science and Technology Policy. "Science and Technology: Shaping the Twenty-First Century, Washington, DC: Executive Office of the President, 1998.

Office of Technology Policy. *Effective Partnering: A Report to Congress on Federal Technology Partnerships*, Washington, DC: U.S. Department of Commerce, 1996.

Olson, Kristen. "Total Science and Engineering Graduate Enrollment Falls for Fourth Consecutive Year," *Data Brief*, Arlington: National Science Foundation, December 17, 1998.

Pakes, Ariel. "On Patents, R&D, and the Stock Market Rate of Return," *Journal of Political Economy*, 93, 1985, pp. 390-409.

Pakes, Ariel. "Patents as Options: Some Estimates of the Value of Holding European Patent Stocks," *Econometrica*, 1986, pp. 755-784.

Pakes, Ariel and Zvi Griliches. "Patents and R&D at the Firm Level: A First Report," *Economics Letters*, 1980 (5), pp. 377-381.

Pakes, Ariel and Zvi Griliches. "Patents and R&D at the Firm Level: A First Look," in *R&D, Patents, and Productivity* (edited by Z. Griliches), Chicago: University of Chicago Press, 1984, pp. 55-72.

Pakes, Ariel and Mark Schankerman. "The Rate of Obsolescence of Patents, Research Gestation Lags, and the Private Rate of Return to Research Outcomes," *R&D, Patents, and Productivity* (edited by Z. Griliches), Chicago: University of Chicago Press, pp. 73-88.

Porter, Michael. *Clusters of Innovation: Regional Foundations of U.S. Competitiveness*. Washington, DC: Council on Competitiveness, 2001.

Prevezer, Martha. "The Dynamics of Industrial Clustering in Biotechnology," *Small Business Economics*, 1997, pp. 255-271.

Reis, Paula and D.H. Thurgood. *Summary Report 1991: Doctoral Recipients from United States Universities*, Washington DC: National Academy Press.

Romeo, Anthony. "Interindustry and Interfirm Differences in the Rate of Diffusion of an Innovation," *Review of Economics and Statistics*, 1975, pp. 311-319.

Romer, Paul M. "Should the Government Subsidize Supply or Demand in the Market of Scientists and Engineers?" NBER Working Paper No. 7723, May 2000.

Rosenberg, Nathan and Richard R. Nelson. "American Universities and Technical Advance in Industry," *Research Policy*, 1994, pp. 323-348.

Saxenian, Annalle. "Regional Networks and the Resurgence of Silicon Valley," *California Management Review*, 1990, pp. 89-111.

Schankerman, Mark and Ariel Pakes. "The Rate of Obsolescence and the Distribution of Patent Values: Some Evidence from European Patent Renewals," *Revue Economique*, 1985, pp. 917-941.

Schankerman, Mark and Ariel Pakes. "Estimates of the Value of Patent Rights in European Countries During the Post-1950 Period," *Economic Journal*, 1986, 96, pp. 1052-1076.

Scherer, F. M. "Firm Size, Market Structure, Opportunity and the Output of Patented Inventions," *American Economic Review*, 1965, pp. 1097-1125.

Scherer, F. M. "Inter-Industry Technology Flows and Productivity Growth," *Review of Economics and Statistics*, 1982, pp. 627-634.

Scherer, F. M. "The Propensity to Patent," *International Journal of Industrial Organization*, 1983, pp. 107-128.

Schmookler, Jacob. *Invention and Economic Growth*, Cambridge, MA: Harvard University Press, 1966.

Schumpeter, Joseph A. "The Instability of Capitalism," *Economic Journal*, 1928, pp. 361-386.

Schumpeter, Joseph A. *Business Cycles*, New York: McGraw Hill, 1939.

Schwalbach, J. and Klaus F. Zimmermann. "A Poisson Model of Patenting and Firm Structure in Germany," in *Innovation and Technological Change: An International Comparison* (edited by Z. Acs and D. B. Audretsch), Ann Arbor: University of Michigan Press, 1991, pp. 109-120.

Scott, John T. "Financing and Leveraging Public/Private Partnerships: The Hurdle-Lowering Auction," *STI Review*, 1998, pp. 67-84.

Siegel, Donald S. "The Impact of Computers on Manufacturing Productivity Growth: A Multiple-Indicators, Multiple-Causes Approach," *Review of Economics and Statistics*, 1997, pp. 68-78.

Siegel, Donald S. *Skill-Biased Technological Change: Evidence from A Firm-Level Survey*, Kalamazoo, MI: W.E. Upjohn Institute Press, 1999.

Siegel, Donald S. and Zvi Griliches. "Purchased Services, Outsourcing, Computers, and Productivity in Manufacturing," in *Output Measurement in the Service Sector,* Chicago: University of Chicago Press, 1992, pp. 429-458.

Siegel, Donald S., Jerry G. Thursby, Marie C. Thursby, and Arvids A. Ziedonis. "Organizational Issues in University-Industry Technology Transfer: An Overview of the Symposium Issue," *Journal of Technology Transfer*, 2001, pp. 5-11.

Siegel, Donald S., David A. Waldman, and Albert N. Link, "Assessing the Impact of Organizational Practices on the Productivity of University Technology Transfer Offices: An Exploratory Study," NBER Working Paper No. 7256, July 1999.

Siegel, Donald S., David A. Waldman, and Albert N. Link. "Assessing the Impact of Organizational Practices on the Relative Productivity of

University Technology Transfer Offices: An Exploratory Study," *Research Policy*, 2003, pp. 27-48.

Sobel, Dava. *Longitude*. New York: Penguin Books, 1995.

Solow, Robert M. "Technical Change and the Aggregate Production Function," *Review of Economics and Statistics*, 1957, pp. 312-320.

Solow, Robert M. "Some Recent Developments in the Theory of Production," in *The Theory and Empirical Analysis of Production* (edited by M. Brown), Princeton: National Bureau of Economic Research, 1967.

Steelman, John R. *Science and Public Policy: A Program for the Nation*. Washington, DC: The U.S. Presidents Scientific Research Board, August 27, 1947.

Stephan, Paula E. "The Economics of Science," *Journal of Economic Literature*, 1996, pp. 1199-1235.

Stephan, Paula E. and Sharon G. Levin. "Exceptional Contributions to U.S. Science by the Foreign-Born and Foreign-Educated," *Population Policy and Research Review*, 2001, pp. 59-79.

Swann, G.M. Peter, Martha Prevezer, and David Stout. *The Dynamics of Industrial Clustering*, Oxford: Oxford University Press, 1998.

Tassey, Gregory. *Technology Infrastructure and Competitive Position*, Norwell, MA: Kluwer Academic Publishers, 1992.

Tassey, Gregory. *The Economics of R&D Policy*, Westport, CT: Quorum Books, 1997.

Tassey, Gregory. "R&D Trends in the U.S. Economy: Strategies and Policy Implications," Gaithersburg, MD: National Institute of Standards and Technology, 1999.

Tassey, Gregory. "Standardization in Technology-Based Markets," *Research Policy*, 2000, pp. 587-602.

Tassey, Gregory. "Underinvestment in Public Good Technologies," *Journal of Technology Transfer*, 2005, pp. 89-113.

Teece, David J. "Economies of Scope and the Scope of the Enterprise, *Journal of Economic Behavior and Organization*, 1980, pp. 223-247.

Terleckyj, Nestor E. *Effects of R&D on the Productivity Growth of Industries: An Explanatory Study*, Washington, DC: National Planning Association, 1974.

Thünen, J.H. von. *The Isolated State in Relation to Agriculture and Political Economy*, Chicago: Loyola University Press, ,1960.

Tibbetts, Ronald, "The Small Business Innovation Research Program and NSF SBIR Commercialization Results," mimeograph, 1999.

Trajtenberg, Manuel. "A Penny for Your Quotes: Patent Citations and the Value of Innovations," *Rand Journal of Economics,* 1990a, pp. 172-187.

Trajtenberg, Manuel. *Economic Analysis of Product Innovation: The Case of CT Scanners,* Cambridge, MA: Harvard University Press, 1990b.

Unesco. *National Science Policies of the U.S.A.: Origins, Development and Present Status.* Paris: Unesco, 1968.

U.S. Bureau of Labor Statistics. www.bls.gov.

U.S. Congress, Office of Technology Assessment. *The Effectiveness of Research and Experimentation Tax Credits,* Washington, DC: Office of Technology Assessment, 1996.

U.S. Department of Commerce. "Emerging Technologies: A Survey of Technical and Economic Opportunities," Washington, DC: Technology Administration, 1990.

U.S. General Accounting Office (GAO). *Managing for Results: Opportunities for Continued Improvements in Agencies' Performance Plans,* Washington, DC: General Accounting Office, 1999.

Usher, A.P. *A History of Mechanical Inventions,* Cambridge: Harvard University Press, 1954.

Wallsten, Scott. "An Empirical Test of Geographic Knowledge Spillovers Using Geographic Information Systems and Firm-Level Data," *Regional Science and Urban Economics,* 2001, pp. 571-599.

Watson, Jason D. "A History of the United States Patent Office," www.m-cam.com, 2001

Wessner, Charles W. *The Small Business Innovation Research Program: An Assessment of the Department of Defense Fast Track,* Washington, DC: National Academy Press, 2000.

Wessner, Charles W. *Government-Industry Partnerships for the Development of New Technologies,* Washington, DC: National Academy Press, 2003.

White House Task Force on Women, Minorities, and the Handicapped in Science and Technology. *Changing America: The New Face of Science and Technology,* Washington, DC: The Task Force, 1989.

INDEX

Printed in the United States
by Baker & Taylor Publisher Services